Mastering Probabilistic Graphical Models Using Python

Master probabilistic graphical models by learning through real-world problems and illustrative code examples in Python

Ankur Ankan

Abinash Panda

[PACKT] open source ⚹

PUBLISHING   community experience distilled

BIRMINGHAM - MUMBAI

Mastering Probabilistic Graphical Models Using Python

First published: July 2015

Production reference: 1280715

Published by Packt Publishing Ltd.
Livery Place
35 Livery Street
Birmingham B3 2PB, UK.

ISBN 978-1-78439-468-4

www.packtpub.com

Credits

Authors
Ankur Ankan
Abinash Panda

Reviewers
Matthieu Brucher
Dave (Jing) Tian
Xiao Xiao

Commissioning Editor
Kartikey Pandey

Acquisition Editors
Vivek Anantharaman
Sam Wood

Content Development Editor
Gaurav Sharma

Technical Editors
Ankita Thakur
Chinmay S. Puranik

Copy Editors
Shambhavi Pai
Swati Priya

Project Coordinator
Bijal Patel

Proofreader
Safis Editing

Indexer
Mariammal Chettiyar

Graphics
Disha Haria

Production Coordinator
Nilesh R. Mohite

Cover Work
Nilesh R. Mohite

About the Authors

Ankur Ankan is a BTech graduate from IIT (BHU), Varanasi. He is currently working in the field of data science. He is an open source enthusiast and his major work includes starting pgmpy with four other members. In his free time, he likes to participate in Kaggle competitions.

> I would like to thank all the pgmpy contributors who have helped me in bringing it to its current stable state. Also, I would like to thank my parents for their relentless support in my endeavors.

Abinash Panda is an undergraduate from IIT (BHU), Varanasi, and is currently working as a data scientist. He has been a contributor to open source libraries such as the Shogun machine learning toolbox and pgmpy, which he started writing along with four other members. He spends most of his free time on improving pgmpy and helping new contributors.

> I would like to thank all the pgmpy contributors. Also, I would like to thank my parents for their support. I am also grateful to all my batchmates of electronics engineering, the class of 2014, for motivating me.

About the Reviewers

Matthieu Brucher holds a master's degree from Ecole Supérieure d'Electricité (information, signals, measures), a master of computer science degree from the University of Paris XI, and a PhD in unsupervised manifold learning from the Université de Strasbourg, France. He is currently an HPC software developer at an oil company and works on next-generation reservoir simulation.

Dave (Jing) Tian is a graduate research fellow and a PhD student in the computer and information science and engineering (CISE) department at the University of Florida. He is a founding member of the Sensei center. His research involves system security, embedded systems security, trusted computing, and compilers. He is interested in Linux kernel hacking, compiler hacking, and machine learning. He also spent a year on AI and machine learning and taught Python and operating systems at the University of Oregon. Before that, he worked as a software developer in the Linux Control Platform (LCP) group at the Alcatel-Lucent (formerly, Lucent Technologies) R&D department for around 4 years. He got his bachelor's and master's degrees from EE in China. He can be reached via his blog at http://davejingtian.org and can be e-mailed at root@davejingtian.org.

Thanks to the authors of this book for doing a good job. I would also like to thank the editors of this book for making it perfect and giving me the opportunity to review such a nice book.

Xiao Xiao got her master's degree from the University of Oregon in 2014. Her research interest lies in probabilistic graphical models. Her previous project was to use probabilistic graphical models to predict human behavior to help people lose weight. Now, Xiao is working as a full-stack software engineer at Poshmark. She was also the reviewer of *Building Probabilistic Graphical Models with Python*, *Packt Publishing*.

www.PacktPub.com

Support files, eBooks, discount offers, and more

For support files and downloads related to your book, please visit www.PacktPub.com.

Did you know that Packt offers eBook versions of every book published, with PDF and ePub files available? You can upgrade to the eBook version at www.PacktPub.com and as a print book customer, you are entitled to a discount on the eBook copy. Get in touch with us at service@packtpub.com for more details.

At www.PacktPub.com, you can also read a collection of free technical articles, sign up for a range of free newsletters and receive exclusive discounts and offers on Packt books and eBooks.

https://www2.packtpub.com/books/subscription/packtlib

Do you need instant solutions to your IT questions? PacktLib is Packt's online digital book library. Here, you can search, access, and read Packt's entire library of books.

Why subscribe?

- Fully searchable across every book published by Packt
- Copy and paste, print, and bookmark content
- On demand and accessible via a web browser

Free access for Packt account holders

If you have an account with Packt at www.PacktPub.com, you can use this to access PacktLib today and view 9 entirely free books. Simply use your login credentials for immediate access.

Table of Contents

Preface   vii

Chapter 1: Bayesian Network Fundamentals   1

Probability theory   2
Random variable   2
Independence and conditional independence   3
Installing tools   5
IPython   5
pgmpy   5
Representing independencies using pgmpy   6
Representing joint probability distributions using pgmpy   7
Conditional probability distribution   8
Representing CPDs using pgmpy   9
Graph theory   11
Nodes and edges   11
Walk, paths, and trails   12
Bayesian models   13
Representation   14
Factorization of a distribution over a network   16
Implementing Bayesian networks using pgmpy   17
Bayesian model representation   18
Reasoning pattern in Bayesian networks   20
D-separation   22
Direct connection   22
Indirect connection   22
Relating graphs and distributions   24
IMAP   24
IMAP to factorization   25
CPD representations   26
Deterministic CPDs   26

Context-specific CPDs	28
Tree CPD	28
Rule CPD	30
Summary	**30**
Chapter 2: Markov Network Fundamentals	**31**
Introducing the Markov network	**32**
Parameterizing a Markov network – factor	33
Factor operations	35
Gibbs distributions and Markov networks	38
The factor graph	**42**
Independencies in Markov networks	**44**
Constructing graphs from distributions	**46**
Bayesian and Markov networks	**47**
Converting Bayesian models into Markov models	47
Converting Markov models into Bayesian models	51
Chordal graphs	53
Summary	**55**
Chapter 3: Inference – Asking Questions to Models	**57**
Inference	**57**
Complexity of inference	59
Variable elimination	**60**
Analysis of variable elimination	66
Finding elimination ordering	69
Using the chordal graph property of induced graphs	71
Minimum fill/size/weight/search	71
Belief propagation	**72**
Clique tree	72
Constructing a clique tree	73
Message passing	76
Clique tree calibration	80
Message passing with division	82
Factor division	83
Querying variables that are not in the same cluster	88
MAP using variable elimination	**90**
Factor maximization	**91**
MAP using belief propagation	**95**
Finding the most probable assignment	**96**
Predictions from the model using pgmpy	**97**
A comparison of variable elimination and belief propagation	**100**
Summary	**101**

Chapter 4: Approximate Inference — 103

The optimization problem — 104
The energy function — 106
Exact inference as an optimization — 107
The propagation-based approximation algorithm — 110
  Cluster graph belief propagation — 112
  Constructing cluster graphs — 115
    Pairwise Markov networks — 115
    Bethe cluster graph — 116
The propagation with approximate messages — 117
  Message creation — 120
  Inference with approximate messages — 123
    Sum-product expectation propagation — 123
    Belief update propagation — 132
Sampling-based approximate methods — 138
Forward sampling — 139
Conditional probability distribution — 141
Likelihood weighting and importance sampling — 141
Importance sampling — 142
Importance sampling in Bayesian networks — 145
  Computing marginal probabilities — 147
  Ratio likelihood weighting — 147
  Normalized likelihood weighting — 147
Markov chain Monte Carlo methods — 148
Gibbs sampling — 148
  Markov chains — 149
The multiple transitioning model — 152
Using a Markov chain — 152
Collapsed particles — 154
Collapsed importance sampling — 155
Summary — 158

Chapter 5: Model Learning – Parameter Estimation in Bayesian Networks — 159

General ideas in learning — 160
  The goals of learning — 160
  Density estimation — 160
  Predicting the specific probability values — 162
  Knowledge discovery — 163
Learning as an optimization — 163
  Empirical risk and overfitting — 164

Discriminative versus generative training **165**
  Learning task 165
    Model constraints 165
    Data observability 166
  Parameter learning **166**
  Maximum likelihood estimation 166
  Maximum likelihood principle 169
  The maximum likelihood estimate for Bayesian networks 171
  Bayesian parameter estimation **175**
  Priors 177
  Bayesian parameter estimation for Bayesian networks 179
  Structure learning in Bayesian networks **183**
  Methods for the learning structure 184
  Constraint-based structure learning 185
    Structure score learning 187
    The likelihood score 187
    The Bayesian score 190
  The Bayesian score for Bayesian networks **193**
  Summary **196**
Chapter 6: Model Learning – Parameter Estimation in Markov Networks **197**
  Maximum likelihood parameter estimation **197**
    Likelihood function 198
      Log-linear model 200
      Gradient ascent 202
    Learning with approximate inference 207
      Belief propagation and pseudo-moment matching 208
    Structure learning 210
      Constraint-based structure learning 210
      Score-based structure learning 212
      The likelihood score 213
      Bayesian score 214
  Summary **216**
Chapter 7: Specialized Models **217**
  The Naive Bayes model **217**
    Why does it even work? 220
    Types of Naive Bayes models 223
      Multivariate Bernoulli Naive Bayes model 224
      Multinomial Naive Bayes model 229
      Choosing the right model 231
  Dynamic Bayesian networks **231**
    Assumptions 231
      Discrete timeline assumption 232

The Markov assumption   232
Model representation   233

The Hidden Markov model   **235**

Generating an observation sequence   238

Computing the probability of an observation   242
The forward-backward algorithm   243
Computing the state sequence   247

Applications   **251**

The acoustic model   252

The language model   253

Summary   **254**

Index   **255**

Preface

This book focuses on the theoretical as well as practical uses of probabilistic graphical models, commonly known as PGM. This is a technique in machine learning in which we use the probability distribution over different variables to learn the model. In this book, we have discussed the different types of networks that can be constructed and the various algorithms for doing inference or predictions over these models. We have added examples wherever possible to make the concepts easier to understand. We also have code examples to promote understanding the concepts more effectively and working on real-life problems.

What this book covers

Chapter 1, Bayesian Network Fundamentals, discusses Bayesian networks (a type of graphical model), its representation, and the independence conditions that this type of network implies.

Chapter 2, Markov Network Fundamentals, discusses the other type of graphical model known as Markov network, its representation, and the independence conditions implied by it.

Chapter 3, Inference – Asking Questions to Models, discusses the various exact inference techniques used in graphical models to predict over newer data points.

Chapter 4, Approximate Inference, discusses the various methods for doing approximate inference in graphical models. As doing exact inference in the case of many real-life problems is computationally very expensive, approximate methods give us a faster way to do inference in such problems.

Chapter 5, Model Learning – Parameter Estimation in Bayesian Networks, discusses the various methods to learn a Bayesian network using data points that we have observed. This chapter also discusses the various methods of learning the network structure with observed data.

Chapter 6, Model Learning – Parameter Estimation in Markov Networks, discusses various methods for learning parameters and network structure in the case of Markov networks.

Chapter 7, Specialized Models, discusses some special cases in Bayesian and Markov models that are very widely used in real-life problems, such as Naive Bayes, Hidden Markov models, and others.

What you need for this book

In this book, we have used IPython to run all the code examples. It is not necessary to use IPython but we recommend you to use it. Most of the code examples use pgmpy and sckit-learn. Also, we have used NumPy at places to generate random data.

Who this book is for

This book will be useful for researchers, machine learning enthusiasts, and people who are working in the data science field and have a basic idea of machine learning or graphical models. This book will help readers to understand the details of graphical models and use them in their day-to-day data science problems.

Conventions

In this book, you will find a number of text styles that distinguish between different kinds of information. Here are some examples of these styles and an explanation of their meaning.

Code words in text, database table names, folder names, filenames, file extensions, pathnames, dummy URLs, user input, and Twitter handles are shown as follows: "We are provided with five variables, namely `sepallength`, `sepalwidth`, `petallength`, `petalwidth`, and `flowerspecies`."

A block of code is set as follows:

```
[default]
raw_data = np.random.randint(low=0, high=2, size=(1000, 5))
data = pd.DataFrame(raw_data, columns=['D', 'I', 'G', 'S', 'L'])

student_model = BayesianModel([('D', 'G'), ('I', 'G'), ('G', 'L'),
('I', 'S')])
```

When we wish to draw your attention to a particular part of a code block, the relevant lines or items are set in bold:

```
[default]
raw_data = np.random.randint(low=0, high=2, size=(1000, 5))
data = pd.DataFrame(raw_data, columns=['D', 'I', 'G', 'S', 'L'])

student_model = BayesianModel([('D', 'G'), ('I', 'G'), ('G', 'L'),
('I', 'S')])

student_model = BayesianModel([('D', 'G'), ('I', 'G'), ('G', 'L'),
('I', 'S')])
```

New terms and **important words** are shown in bold.

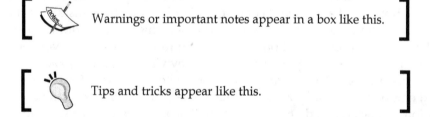

> Warnings or important notes appear in a box like this.

> Tips and tricks appear like this.

Reader feedback

Feedback from our readers is always welcome. Let us know what you think about this book—what you liked or disliked. Reader feedback is important for us as it helps us develop titles that you will really get the most out of.

To send us general feedback, simply e-mail feedback@packtpub.com, and mention the book's title in the subject of your message.

If there is a topic that you have expertise in and you are interested in either writing or contributing to a book, see our author guide at www.packtpub.com/authors.

Customer support

Now that you are the proud owner of a Packt book, we have a number of things to help you to get the most from your purchase.

Downloading the example code

You can download the example code files from your account at `http://www.packtpub.com` for all the Packt Publishing books you have purchased. If you purchased this book elsewhere, you can visit `http://www.packtpub.com/support` and register to have the files e-mailed directly to you.

Downloading the color images of this book

We also provide you with a PDF file that has color images of the screenshots/diagrams used in this book. The color images will help you better understand the changes in the output. You can download this file from `http://www.packtpub.com/sites/default/files/downloads/4684OS_ColorImages.pdf`.

Errata

Although we have taken every care to ensure the accuracy of our content, mistakes do happen. If you find a mistake in one of our books—maybe a mistake in the text or the code—we would be grateful if you could report this to us. By doing so, you can save other readers from frustration and help us improve subsequent versions of this book. If you find any errata, please report them by visiting `http://www.packtpub.com/submit-errata`, selecting your book, clicking on the **Errata Submission Form** link, and entering the details of your errata. Once your errata are verified, your submission will be accepted and the errata will be uploaded to our website or added to any list of existing errata under the Errata section of that title.

To view the previously submitted errata, go to `https://www.packtpub.com/books/content/support` and enter the name of the book in the search field. The required information will appear under the **Errata** section.

Piracy

Piracy of copyrighted material on the Internet is an ongoing problem across all media. At Packt, we take the protection of our copyright and licenses very seriously. If you come across any illegal copies of our works in any form on the Internet, please provide us with the location address or website name immediately so that we can pursue a remedy.

Please contact us at copyright@packtpub.com with a link to the suspected pirated material.

We appreciate your help in protecting our authors and our ability to bring you valuable content.

Questions

If you have a problem with any aspect of this book, you can contact us at questions@packtpub.com, and we will do our best to address the problem.

1
Bayesian Network Fundamentals

A **graphical model** is essentially a way of representing joint probability distribution over a set of random variables in a compact and intuitive form. There are two main types of graphical models, namely **directed** and **undirected**. We generally use a directed model, also known as a **Bayesian network**, when we mostly have a causal relationship between the random variables. Graphical models also give us tools to operate on these models to find conditional and marginal probabilities of variables, while keeping the computational complexity under control.

In this chapter, we will cover:

- The basics of random variables, probability theory, and graph theory
- Bayesian models
- Independencies in Bayesian models
- The relation between graph structure and probability distribution in Bayesian networks (IMAP)
- Different ways of representing a conditional probability distribution
- Code examples for all of these using `pgmpy`

Probability theory

To understand the concepts of probability theory, let's start with a real-life situation. Let's assume we want to go for an outing on a weekend. There are a lot of things to consider before going: the weather conditions, the traffic, and many other factors. If the weather is windy or cloudy, then it is probably not a good idea to go out. However, even if we have information about the weather, we cannot be completely sure whether to go or not; hence we have used the words probably or maybe. Similarly, if it is windy in the morning (or at the time we took our observations), we cannot be completely certain that it will be windy throughout the day. The same holds for cloudy weather; it might turn out to be a very pleasant day. Further, we are not completely certain of our observations. There are always some limitations in our ability to observe; sometimes, these observations could even be noisy. In short, *uncertainty* or *randomness* is the innate nature of the world. The **probability theory** provides us the necessary tools to study this uncertainty. It helps us look into options that are unlikely yet probable.

Random variable

Probability deals with the study of events. From our intuition, we can say that some events are more likely than others, but to quantify the likeliness of a particular event, we require the *probability theory*. It helps us predict the future by assessing how likely the outcomes are.

Before going deeper into the probability theory, let's first get acquainted with the basic terminologies and definitions of the probability theory. A random variable is a way of representing an attribute of the outcome. Formally, a random variable X is a function that maps a possible set of outcomes Ω to some set E, which is represented as follows:

$$X : \Omega \to E$$

As an example, let us consider the *outing* example again. To decide whether to go or not, we may consider the skycover (to check whether it is cloudy or not). Skycover is an attribute of the day. Mathematically, the random variable *skycover* (X) is interpreted as a function, which maps the day (Ω) to its *skycover values (E)*. So when we say the event $X = 40.1$, it represents the set of all the days $\{\omega\}$ such that $f_{skycover}(w) = 40.1$, where $f_{skycover}$ is the mapping function. Formally speaking, $\{w \epsilon \Omega : f_{skycover}(w) = 40.1\}$.

Random variables can either be *discrete* or *continuous*. A discrete random variable can only take a finite number of values. For example, the random variable representing the outcome of a coin toss can take only two values, heads or tails; and hence, it is discrete. Whereas, a continuous random variable can take infinite number of values. For example, a variable representing the speed of a car can take any number values.

For any event whose outcome is represented by some random variable *(X)*, we can assign some value to each of the possible outcomes of *X*, which represents how probable it is. This is known as the probability distribution of the random variable and is denoted by *P(X)*.

For example, consider a set of restaurants. Let *X* be a random variable representing the quality of food in a restaurant. It can take up a set of values, such as *{good, bad, average}*. *P(X)*, represents the probability distribution of *X*, that is, if *P(X = good) = 0.3*, *P(X = average) = 0.5*, and *P(X = bad) = 0.2*. This means there is 30 percent chance of a restaurant serving good food, 50 percent chance of it serving average food, and 20 percent chance of it serving bad food.

Independence and conditional independence

In most of the situations, we are rather more interested in looking at multiple attributes at the same time. For example, to choose a restaurant, we won't only be looking just at the quality of food; we might also want to look at other attributes, such as the cost, location, size, and so on. We can have a probability distribution over a combination of these attributes as well. This type of distribution is known as **joint probability distribution**. Going back to our restaurant example, let the random variable for the quality of food be represented by *Q*, and the cost of food be represented by *C*. *Q* can have three categorical values, namely *{good, average, bad}*, and *C* can have the values *{high, low}*. So, the joint distribution for *P(Q, C)* would have probability values for all the combinations of states of *Q* and *C*. *P(Q = good, C = high)* will represent the probability of a pricey restaurant with good quality food, while *P(Q = bad, C = low)* will represent the probability of a restaurant that is less expensive with bad quality food.

Let us consider another random variable representing an attribute of a restaurant, its location *L*. The cost of food in a restaurant is not only affected by the quality of food but also the location (generally, a restaurant located in a very good location would be more costly as compared to a restaurant present in a not-very-good location). From our intuition, we can say that the probability of a costly restaurant located at a very good location in a city would be different (generally, more) than simply the probability of a costly restaurant, or the probability of a cheap restaurant located at a prime location of city is different (generally less) than simply probability of a cheap restaurant. Formally speaking, *P(C = high | L = good)* will be different from *P(C = high)* and *P(C = low | L = good)* will be different from *P(C = low)*. This indicates that the random variables *C* and *L* are not independent of each other.

These attributes or random variables need not always be dependent on each other. For example, the quality of food doesn't depend upon the location of restaurant. So, $P(Q = good \mid L = good)$ or $P(Q = good \mid L = bad)$ would be the same as $P(Q = good)$, that is, our estimate of the quality of food of the restaurant will not change even if we have knowledge of its location. Hence, these random variables are *independent* of each other.

In general, random variables $\{X_1, X_2, \ldots, X_2\}$ can be considered as *independent* of each other, if:

$$P(X_1, X_2, \ldots, X_n) = P(X_1) P(X_2) \ldots P(X_n)$$

They may also be considered independent if:

$$P(X_1, X_2, \ldots, X_n) = \prod_{i=1}^{n} P(X_i)$$

We can easily derive this conclusion. We know the following from the chain rule of probability:

$$P(X, Y) = P(X)\, P(Y \mid X)$$

If Y is independent of X, that is, if $X \mid Y$, then $P(Y \mid X) = P(Y)$. Then:

$$P(X, Y) = P(X)\, P(Y)$$

Extending this result on multiple variables, we can easily get to the conclusion that a set of random variables are independent of each other, if their joint probability distribution is equal to the product of probabilities of each individual random variable.

Sometimes, the variables might not be independent of each other. To make this clearer, let's add another random variable, that is, the number of people visiting the restaurant N. Let's assume that, from our experience we know the number of people visiting *only* depends on the cost of food at the restaurant and its location (generally, lesser number of people visit costly restaurants). Does the quality of food Q affect the number of people visiting the restaurant? To answer this question, let's look into the random variable affecting N, cost C, and location L. As C is directly affected by Q, we can conclude that Q affects N. However, let's consider a situation when we know that the restaurant is costly, that is, $C = high$ and let's ask the same question, "does the quality of food affect the number of people coming to the restaurant?". The answer is *no*. The number of people coming only depends on the price and location, so if we know that the cost is high, then we can easily conclude that fewer people will visit, irrespective of the quality of food. Hence, $Q \perp N \mid C$.

This type of independence is called **conditional independence**.

Installing tools

Let's now see some coding examples using pgmpy, to represent joint distributions and independencies. Here, we will mostly work with IPython and pgmpy (and a few other libraries) for coding examples. So, before moving ahead, let's get a basic introduction to these.

IPython

IPython is a command shell for interactive computing in multiple programming languages, originally developed for the Python programming language, which offers enhanced introspection, rich media, additional shell syntax, tab completion, and a rich history. IPython provides the following features:

- Powerful interactive shells (terminal and Qt-based)
- A browser-based notebook with support for code, text, mathematical expressions, inline plots, and other rich media
- Support for interactive data visualization and use of GUI toolkits
- Flexible and embeddable interpreters to load into one's own projects
- Easy-to-use and high performance tools for parallel computing

You can install IPython using the following command:

```
>>> pip3 install ipython
```

To start the IPython command shell, you can simply type ipython3 in the terminal. For more installation instructions, you can visit http://ipython.org/install.html.

pgmpy

pgmpy is a Python library to work with Probabilistic Graphical models. As it's currently not on PyPi, we will need to build it manually. You can get the source code from the Git repository using the following command:

```
>>> git clone https://github.com/pgmpy/pgmpy
```

Now cd into the cloned directory switch branch for version used in this book and build it with the following code:

```
>>> cd pgmpy
>>> git checkout book/v0.1
>>> sudo python3 setup.py install
```

For more installation instructions, you can visit `http://pgmpy.org/install.html`.

With both IPython and pgmpy installed, you should now be able to run the examples in the book.

Representing independencies using pgmpy

To represent independencies, pgmpy has two classes, namely `IndependenceAssertion` and `Independencies`. The `IndependenceAssertion` class is used to represent individual assertions of the form of $(X \perp Y)$ or $(X \perp Y | Z)$. Let's see some code to represent assertions:

```
# Firstly we need to import IndependenceAssertion
In [1]: from pgmpy.independencies import IndependenceAssertion
# Each assertion is in the form of [X, Y, Z] meaning X is
# independent of Y given Z.
In [2]: assertion1 = IndependenceAssertion('X', 'Y')
In [3]: assertion1
Out[3]: (X _|_ Y)
```

Here, `assertion1` represents that the variable X is independent of the variable Y. To represent conditional assertions, we just need to add a third argument to `IndependenceAssertion`:

```
In  [4]: assertion2 = IndependenceAssertion('X', 'Y', 'Z')
In  [5]: assertion2
Out [5]: (X _|_ Y | Z)
```

In the preceding example, `assertion2` represents $(X \perp Y | Z)$.

`IndependenceAssertion` also allows us to represent assertions in the form of $(X \perp Y, Z | A, B)$. To do this, we just need to pass a list of random variables as arguments:

```
In [4]: assertion2 = IndependenceAssertion('X', 'Y', 'Z')
In [5]: assertion2
Out[5]: (X _|_ Y | Z)
```

Moving on to the `Independencies` class, an `Independencies` object is used to represent a set of assertions. Often, in the case of Bayesian or Markov networks, we have more than one assertion corresponding to a given model, and to represent these independence assertions for the models, we generally use the `Independencies` object. Let's take a few examples:

```
In [8]: from pgmpy.independencies import Independencies
# There are multiple ways to create an Independencies object, we
# could either initialize an empty object or initialize with some
# assertions.

In [9]: independencies = Independencies() # Empty object
In [10]: independencies.get_assertions()
Out[10]: []

In [11]: independencies.add_assertions(assertion1, assertion2)
In [12]: independencies.get_assertions()
Out[12]: [(X _|_ Y), (X _|_ Y | Z)]
```

We can also directly initialize `Independencies` in these two ways:

```
In [13]: independencies = Independencies(assertion1, assertion2)
In [14]: independencies = Independencies(['X', 'Y'],
                                         ['A', 'B', 'C'])
In [15]: independencies.get_assertions()
Out[15]: [(X _|_ Y), (A _|_ B | C)]
```

Representing joint probability distributions using pgmpy

We can also represent joint probability distributions using pgmpy's `JointProbabilityDistribution` class. Let's say we want to represent the joint distribution over the outcomes of tossing two fair coins. So, in this case, the probability of all the possible outcomes would be 0.25, which is shown as follows:

```
In [16]: from pgmpy.factors import JointProbabilityDistribution as
         Joint
In [17]: distribution = Joint(['coin1', 'coin2'],
                              [2, 2],
                              [0.25, 0.25, 0.25, 0.25])
```

Here, the first argument includes names of random variable. The second argument is a list of the number of states of each random variable. The third argument is a list of probability values, assuming that the first variable changes its states the slowest. So, the preceding distribution represents the following:

```
In [18]: print(distribution)
```

coin1	coin2	P(coin1,coin2)
coin1_0	coin2_0	0.2500
coin1_0	coin2_1	0.2500
coin1_1	coin2_0	0.2500
coin1_1	coin2_1	0.2500

We can also conduct independence queries over these distributions in pgmpy:

```
In [19]: distribution.check_independence('coin1', 'coin2')
Out[20]: True
```

Conditional probability distribution

Let's take an example to understand conditional probability better. Let's say we have a bag containing three apples and five oranges, and we want to randomly take out fruits from the bag one at a time without replacing them. Also, the random variables X_1 and X_2 represent the outcomes in the first try and second try respectively. So, as there are three apples and five oranges in the bag initially, $P(X_1 = apple) = 0.375$ and $P(X_1 = orange) = 0.625$. Now, let's say that in our first attempt we got an orange. Now, we cannot simply represent the probability of getting an apple or orange in our second attempt. The probabilities in the second attempt will depend on the outcome of our first attempt and therefore, we use conditional probability to represent such cases. Now, in the second attempt, we will have the following probabilities that depend on the outcome of our first try: $P(X_2 = apple | X_1 = orange) = \frac{3}{7}$, $P(X_2 = orange | X_1 = orange) = \frac{4}{7}$, $P(X_2 = apple | X_1 = apple) = \frac{2}{7}$, and $P(X_2 = orange | X_1 = apple) = \frac{5}{7}$.

The **Conditional Probability Distribution (CPD)** of two variables X_1 and X_2 can be represented as $P(X_1 | X_2)$, representing the probability of X_1 given X_2 that is the probability of X_1 after the event X_2 has occurred and we know it's outcome. Similarly, we can have $P(X_2 | X_1)$ representing the probability of X_2 after having an observation for X_1.

The simplest representation of CPD is tabular CPD. In a tabular CPD, we construct a table containing all the possible combinations of different states of the random variables and the probabilities corresponding to these states. Let's consider the earlier restaurant example.

Let's begin by representing the marginal distribution of the quality of food with Q. As we mentioned earlier, it can be categorized into three values {*good, bad, average*}. For example, $P(Q)$ can be represented in the tabular form as follows:

Quality	P(Q)
Good	0.3
Normal	0.5
Bad	0.2

Similarly, let's say $P(L)$ is the probability distribution of the location of the restaurant. Its CPD can be represented as follows:

Location	P(L)
Good	0.6
Bad	0.4

As the cost of restaurant C depends on both the quality of food Q and its location L, we will be considering $P(C \mid Q, L)$, which is the conditional distribution of C, given Q and L:

Location	Good			Bad		
Quality	Good	Normal	Bad	Good	Normal	Bad
Cost						
High	0.8	0.6	0.1	0.6	0.6	0.05
Low	0.2	0.4	0.9	0.4	0.4	0.95

Representing CPDs using pgmpy

Let's first see how to represent the tabular CPD using pgmpy for variables that have no conditional variables:

```
In [1]: from pgmpy.factors import TabularCPD

# For creating a TabularCPD object we need to pass three
# arguments: the variable name, its cardinality that is the number
# of states of the random variable and the probability value
```

```
# corresponding each state.
In [2]: quality = TabularCPD(variable='Quality',
                             variable_card=3,
                             values=[[0.3], [0.5], [0.2]])

In [3]: print(quality)
```

['Quality', 0]	0.3
['Quality', 1]	0.5
['Quality', 2]	0.2

```
In [4]: quality.variables
Out[4]: OrderedDict([('Quality', [State(var='Quality', state=0),
                                  State(var='Quality', state=1),
                                  State(var='Quality', state=2)])])

In [5]: quality.cardinality
Out[5]: array([3])

In [6]: quality.values
Out[6]: array([0.3, 0.5, 0.2])
```

You can see here that the values of the CPD are a 1D array instead of a 2D array, which you passed as an argument. Actually, pgmpy internally stores the values of the TabularCPD as a flattened numpy array. We will see the reason for this in the next chapter.

```
In [7]: location = TabularCPD(variable='Location',
                              variable_card=2,
                              values=[[0.6], [0.4]])

In [8]: print(location)
```

['Location', 0]	0.6
['Location', 1]	0.4

However, when we have conditional variables, we also need to specify them and the cardinality of those variables. Let's define the TabularCPD for the cost variable:

```
In [9]: cost = TabularCPD(
                    variable='Cost',
                    variable_card=2,
                    values=[[0.8, 0.6, 0.1, 0.6, 0.6, 0.05],
                            [0.2, 0.4, 0.9, 0.4, 0.4, 0.95]],
                    evidence=['Q', 'L'],
                    evidence_card=[3, 2])
```

Graph theory

The second major framework for the study of probabilistic graphical models is graph theory. Graphs are the skeleton of PGMs, and are used to compactly encode the independence conditions of a probability distribution.

Nodes and edges

The foundation of graph theory was laid by Leonhard Euler when he solved the famous *Seven Bridges of Konigsberg* problem. The city of Konigsberg was set on both sides by the Pregel river and included two islands that were connected and maintained by seven bridges. The problem was to find a walk to exactly cross all the bridges once in a single walk.

To visualize the problem, let's think of the graph in Fig 1.1:

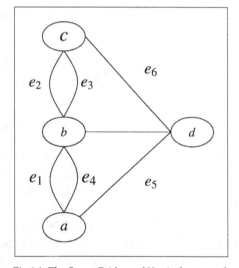

Fig 1.1: The Seven Bridges of Konigsberg graph

Here, the nodes **a**, **b**, **c**, and **d** represent the land, and are known as **vertices** of the graph. The line segments *ab*, *bc*, *cd*, *da*, *ab*, and *bc* connecting the land parts are the bridges and are known as the edges of the graph. So, we can think of the problem of crossing all the bridges once in a single walk as tracing along all the edges of the graph without lifting our pencils.

Formally, a graph $G = (V, E)$ is an ordered pair of finite sets. The elements of the set V are known as the nodes or the vertices of the graph, and the elements of $E \subseteq |V|^2$ are the edges or the arcs of the graph. The number of nodes or cardinality of G, denoted by $|V|$, are known as the order of the graph. Similarly, the number of edges denoted by $|E|$ are known as the size of the graph. Here, we can see that the Konigsberg city graph shown in Fig 1.1 is of order 4 and size 7.

In a graph, we say that two vertices, $u, v \in V$ are adjacent if $u, v \in E$. In the City graph, all the four vertices are adjacent to each other because there is an edge for every possible combination of two vertices in the graph. Also, for a vertex $v \in V$, we define the neighbors set of v as $(u, v) \in E$. In the City graph, we can see that b and d are neighbors of c. Similarly, a, b, and c are neighbors of d.

We define an edge to be a self loop if the start vertex and the end vertex of the edge are the same. We can put it more formally as, any edge of the form (u, u), where $u \in V$ is a self loop.

Until now, we have been talking only about graphs whose edges don't have a direction associated with them, which means that the edge *(u, v)* is same as the edge *(v, u)*. These types of graphs are known as undirected graphs. Similarly, we can think of a graph whose edges have a sense of direction associated with it. For these graphs, the edge set E would be a set of ordered pair of vertices. These types of graphs are known as directed graphs. In the case of a directed graph, we also define the indegree and outdegree for a vertex. For a vertex $v \in V$, we define its outdegree as the number of edges originating from the vertex v, that is, $\left| \{ u \mid (v, u) \in E \} \right|$. Similarly, the indegree is defined as the number of edges that end at the vertex v, that is, $\left| \{ u \mid (u, v) \in E \} \right|$.

Walk, paths, and trails

For a graph $G = (V, E)$ and $u, v \in V$, we define a u - v walk as an alternating sequence of vertices and edges, starting with u and ending with v. In the City graph of Fig 1.1, we can have an example of a - d walk as $W : a, e_1, b, e_2, c, e_3, b, e_6, d$.

If there aren't multiple edges between the same vertices, then we simply represent a walk by a sequence of vertices. As in the case of the Butterfly graph shown in Fig 1.2, we can have a walk $W : a, c, d, c, e$:

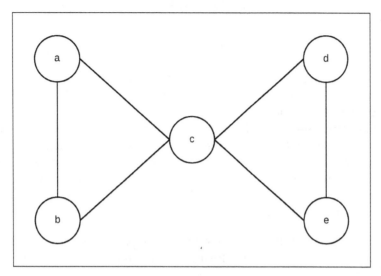

Fig 1.2: Butterfly graph — a undirected graph

A walk with no repeated edges is known as a trail. For example, the walk $W : a,e_1,b,e_2,c,e_3,b,e_4,a$ in the City graph is a trail. Also, a walk with no repeated vertices, except possibly the first and the last, is known as a path. For example, the walk $W : a,e_1,b,e_2,c,e_7,d,e_5,a$ in the City graph is a path.

Also, a graph is known as *cyclic* if there are one or more paths that start and end at the same node. Such paths are known as **cycles**. Similarly, if there are no cycles in a graph, it is known as an acyclic graph.

Bayesian models

In most of the real-life cases when we would be representing or modeling some event, we would be dealing with a lot of random variables. Even if we would consider all the random variables to be discrete, there would still be exponentially large number of values in the joint probability distribution. Dealing with such huge amount of data would be computationally expensive (and in some cases, even intractable), and would also require huge amount of memory to store the probability of each combination of states of these random variables.

However, in most of the cases, many of these variables are marginally or conditionally independent of each other. By exploiting these independencies, we can reduce the number of values we need to store to represent the joint probability distribution.

For instance, in the previous restaurant example, the joint probability distribution across the four random variables that we discussed (that is, quality of food Q, location of restaurant L, cost of food C, and the number of people visiting N) would require us to store 23 independent values. By the chain rule of probability, we know the following:

$$P(Q, L, C, N) = P(Q) \, P(L \mid Q) \, P(C \mid L, Q) \, P(N \mid C, Q, L)$$

Now, let us try to exploit the marginal and conditional independence between the variables, to make the representation more compact. Let's start by considering the independency between the location of the restaurant and quality of food over there. As both of these attributes are independent of each other, $P(L \mid Q)$ would be the same as $P(L)$. Therefore, we need to store only one parameter to represent it. From the conditional independence that we have seen earlier, we know that $N \perp Q \mid C$. Thus, $P(N \mid C, Q, L)$ would be the same as $P(N \mid C, L)$; thus needing only four parameters. Therefore, we now need only ($2 + 1 + 6 + 4 = 13$) parameters to represent the whole distribution.

We can conclude that exploiting independencies helps in the compact representation of joint probability distribution. This forms the basis for the Bayesian network.

Representation

A Bayesian network is represented by a **Directed Acyclic Graph (DAG)** and a set of **Conditional Probability Distributions (CPD)** in which:

- The nodes represent random variables
- The edges represent dependencies
- For each of the nodes, we have a CPD

In our previous restaurant example, the nodes would be as follows:

- Quality of food (Q)
- Location (L)
- Cost of food (C)
- Number of people (N)

As the cost of food was dependent on the quality of food (Q) and the location of the restaurant (L), there will be an edge each from $Q \rightarrow C$ and $L \rightarrow C$. Similarly, as the number of people visiting the restaurant depends on the price of food and its location, there would be an edge each from $L \rightarrow N$ and $C \rightarrow N$. The resulting structure of our Bayesian network is shown in Fig 1.3:

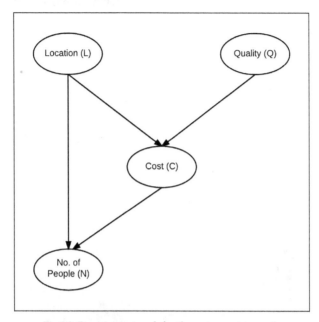

Fig 1.3: Bayesian network for the restaurant example

Factorization of a distribution over a network

Each node in our Bayesian network for restaurants has a CPD associated to it. For example, the CPD for the cost of food in the restaurant is $P(C \mid Q, L)$, as it only depends on the quality of food and location. For the number of people, it would be $P(N \mid C, L)$. So, we can generalize that the CPD associated with each node would be $P(node \mid Par(node))$ where $Par(node)$ denotes the parents of the node in the graph. Assuming some probability values, we will finally get a network as shown in Fig 1.4:

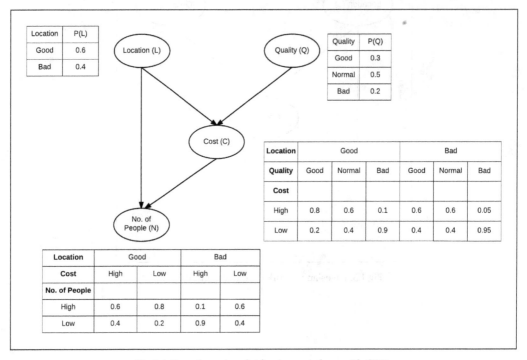

Fig 1.4: Bayesian network of restaurant along with CPDs

Let us go back to the joint probability distribution of all these attributes of the restaurant again. Considering the independencies among variables, we concluded as follows:

$$P(Q,C,L,N) = P(Q)P(L)P(C \mid Q, L)P(N \mid C, L)$$

So now, looking into the **Bayesian network (BN)** for the restaurant, we can say that for any Bayesian network, the *joint probability distribution* $P(X_1, X_2, \dots, X_n)$ over all its random variables $\{X_1, X_2, \dots, X_2\}$ can be represented as follows:

$$P(X_1, X_2, \dots, X_n) = \prod_{i=1}^{n} P(X_i \mid Par(X_i))$$

This is known as the chain rule for Bayesian networks.

Also, we say that a distribution *P factorizes* over a graph *G*, if *P* can be encoded as follows:

$$P(X_1, X_2, \ldots, X_n) = \prod_{i=1}^{n} P(X_i \mid Par_G(X_i))$$

Here, $Par_G(X)$ is the parent of *X* in the graph *G*.

Implementing Bayesian networks using pgmpy

Let us consider a more complex *Bayesian network* of a student getting late for school, as shown in Fig 1.5:

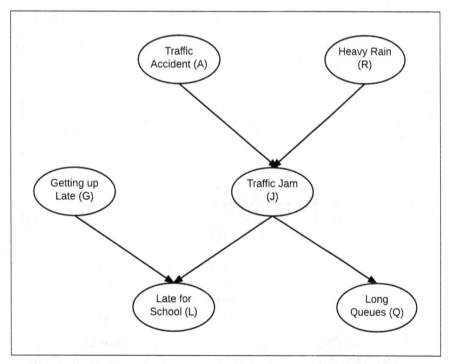

Fig 1.5: Bayesian network representing a particular day of a student going to school

For this *Bayesian network*, just for simplicity, let us assume that each *random variable* is *discrete* with only two possible states *{yes, no}*.

Bayesian model representation

In pgmpy, we can initialize an empty BN or a model with nodes and edges. We can initializing an empty model as follows:

```
In [1]: from pgmpy.models import BayesianModel
In [2]: model = BayesianModel()
```

We can now add nodes and edges to this network:

```
In [3]: model.add_nodes_from(['rain', 'traffic_jam'])
In [4]: model.add_edge('rain', 'traffic_jam')
```

If we add an edge, but the nodes, between which the edge is, are not present in the model, pgmpy automatically adds those nodes to the model.

```
In [5]: model.add_edge('accident', 'traffic_jam')
In [6]: model.nodes()
Out[6]: ['accident', 'rain', 'traffic_jam']
In [7]: model.edges()
Out[7]: [('rain', 'traffic_jam'), ('accident', 'traffic_jam')]
```

In the case of a Bayesian network, each of the nodes has an associated CPD with it. So, let's define some tabular CPDs to associate with the model:

 The name of the variable in tabular CPD should be exactly the same as the name of the node used while creating the Bayesian network, as pgmpy internally uses this name to match the tabular CPDs with the nodes.

```
In [8]: from pgmpy.factors import TabularCPD
In [9]: cpd_rain = TabularCPD('rain', 2, [[0.4], [0.6]])
In [10]: cpd_accident = TabularCPD('accident', 2, [[0.2], [0.8]])
In [11]: cpd_traffic_jam = TabularCPD(
                    'traffic_jam', 2,
                    [[0.9, 0.6, 0.7, 0.1],
                     [0.1, 0.4, 0.3, 0.9]],
                    evidence=['rain', 'accident'],
                    evidence_card=[2, 2])
```

Here, we defined three CPDs. We now need to associate them with our model. To associate them with the model, we just need to use the add_cpd method and pgmpy automatically figures out which CPD is for which node:

```
In [12]: model.add_cpds(cpd_rain, cpd_accident, cpd_traffic_jam)
In [13]: model.get_cpds()
Out[13]:
```

```
[<TabularCPD representing P(rain:2) at 0x7f477b6f9940>,
 <TabularCPD representing P(accident:2) at 0x7f477b6f97f0>,
 <TabularCPD representing P(traffic_jam:2 | rain:2, accident:2) at
                                            0x7f477b6f9e48>]
```

Now, let's add the remaining variables and their CPDs:

```
In [14]: model.add_node('long_queues')
In [15]: model.add_edge('traffic_jam', 'long_queues')
In [16]: cpd_long_queues = TabularCPD('long_queues', 2,
                            [[0.9, 0.2],
                             [0.1, 0.8]],
                            evidence=['traffic_jam'],
                            evidence_card=[2])
In [17]: model.add_cpds(cpd_long_queues)
In [18]: model.add_nodes_from(['getting_up_late',
                         'late_for_school'])
In [19]: model.add_edges_from(
                   [('getting_up_late', 'late_for_school'),
                    ('traffic_jam', 'late_for_school')])
In [20]: cpd_getting_up_late = TabularCPD('getting_up_late', 2,
                                    [[0.6], [0.4]])
In [21]: cpd_late_for_school = TabularCPD(
                            'late_for_school', 2,
                            [[0.9, 0.45, 0.8, 0.1],
                             [0.1, 0.55, 0.2, 0.9]],
                            evidence=['getting_up_late',
                                      'traffic_jam'],
                            evidence_card=[2, 2])
In [22]: model.add_cpds(cpd_getting_up_late, cpd_late_for_school)
In [23]: model.get_cpds()
Out[23]:
[<TabularCPD representing P(rain:2) at 0x7f477b6f9940>,
 <TabularCPD representing P(accident:2) at 0x7f477b6f97f0>,
 <TabularCPD representing P(traffic_jam:2 | rain:2, accident:2) at
                                            0x7f477b6f9e48>,
 <TabularCPD representing P(long_queues:2 | traffic_jam:2) at
                                            0x7f477b7051d0>,
 <TabularCPD representing P(getting_up_late:2) at 0x7f477b7059e8>,
 <TabularCPD representing P(late_for_school:2 | getting_up_late:2,
                            traffic_jam:2) at 0x7f477b705dd8>]
```

Additionally, `pgmpy` also provides a `check_model` method that checks whether the model and all the associated CPDs are consistent:

```
In [24]: model.check_model()
Out[25]: True
```

In case we have got some wrong CPD associated with the model and we want to remove it, we can use the `remove_cpd` method. Let's say we want to remove the CPD associated with variable `late_for_school`, we could simply do as follows:

```
In [26]: model.remove_cpds('late_for_school')
In [27]: model.get_cpds()
Out[27]:
[<TabularCPD representing P(rain:2) at 0x7f477b6f9940>,
 <TabularCPD representing P(accident:2) at 0x7f477b6f97f0>,
 <TabularCPD representing P(traffic_jam:2 | rain:2, accident:2) at
                                         0x7f477b6f9e48>,
 <TabularCPD representing P(long_queues:2 | traffic_jam:2) at
                                         0x7f477b7051d0>,
 <TabularCPD representing P(getting_up_late:2) at 0x7f477b7059e8>]
```

Reasoning pattern in Bayesian networks

Would the probability of having a road accident change if I knew that there was a traffic jam? Or, what are the chances that it rained heavily today if some student comes late to class? *Bayesian networks* helps in finding answers to all these questions. *Reasoning patterns* are key elements of *Bayesian networks*.

Before answering all these questions, we need to compute the joint probability distribution. For ease in naming the nodes, let's denote them as follows:

- Traffic accident as A
- Heavy rain as B
- Traffic jam as J
- Getting up late as G
- Long queues as Q
- Late to school as L

From the chain rule of the Bayesian network, we have the joint probability distribution P_J as follows:

$$P_J = P(A, R, J, G, L, Q) = P(A)P(R)P(J \mid A, R)P(Q \mid G)P(L \mid G, J)$$

Starting with a simple query, what are the chances of having a traffic jam if I know that there was a road accident? This question can be put formally as what is the value of $P(J | A = True)$?

First, let's compute the probability of having a traffic jam $P(J)$. $P(J)$ can be computed by summing all the cases in the joint probability distribution, where $J = True$ and $J = False$, and then renormalize the distribution to sum it to 1. We get $P(J = True) = 0.416$ and $P(J = True) = 0.584$.

To compute $P(J | A = True)$, we have to eliminate all the cases where $A = False$, and then we can follow the earlier procedure to get $P(J | A = True)$. This results in $P(J = True | A = True) = 0.72$ and $P(J = False | A = True) = 0.28$. We can see that the chances of having a traffic jam increased when we knew that there was an accident. These results match with our intuition. From this, we conclude that the observation of the outcome of the parent in a Bayesian network influences the probability of its children. This is known as **causal reasoning**. Causal reasoning need not only be the effect of parent on its children; it can go further downstream in the network.

We have seen that the observation of the outcome of parents influence the probability of the children. Is the inverse possible? Let's try to find the probability of heavy rain if we know that there is a traffic accident. To do so, we have to eliminate all the cases where $J = False$ and then reduce the probability to get $P(R | J = True)$. This results in $P(R = True | J = True) = 0.7115$ and $P(R = False | J = True) = 0.2885$. This is also intuitive. If we knew that there was a traffic jam, then the chances of heavy rain would increase. This is known as *evidential reasoning*, where the observation of the outcomes of the children or effect influences the probability of parents or causes.

Let's look at another type of reasoning pattern. If we knew that there was a traffic jam on a day when there was no heavy rain, would it affect the chances of a traffic accident? To do so, we have to follow a similar procedure of eliminating all those cases, except the ones where $R = False$ and $J = True$. By doing so, we would get $P(A = True | J = True, R = False) = 0.6$ and $P(A = False | J = True, R = False) = 0.4$. Now, the probability of an accident increases, which is what we had expected. As we can see that before the observation of the traffic jam, both the random variables, heavy rain and traffic accident, were independent of each other, but with the observation of their common children, they are now dependent on each other. This type of reasoning is called as *intercausal reasoning*, where different causes with the same effect influence each other.

D-separation

In the last section, we saw how influence flows in a Bayesian network, and how observing some event changes our belief about other variables in the network. In this section, we will discuss the independence conditions that hold in a Bayesian network no matter which probability distribution is parameterizing the network.

In any network, there can be two types of connections between variables, *direct* or *indirect*. Let's start by discussing the direct connection between variables.

Direct connection

In the case of direct connections, we have a direct connection between two variables, that is, there's an edge $X \rightarrow Y$ in the graph G. In the case of a direct connection, we can always find some probability distribution where they are dependent. Hence, there is no independence condition in a direct connection, no matter which other variables are observed in the network.

Indirect connection

In the case of indirect connections, we have four different ways in which the variables can be connected. They are as follows:

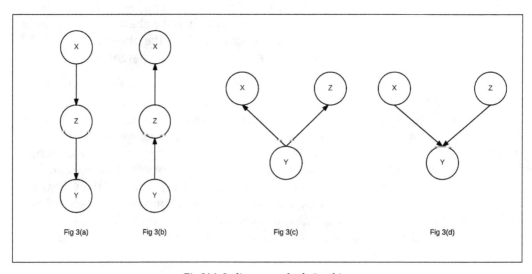

Fig 3(a): Indirect causal relationship

Fig 3(b): Indirect evidential relationship

Fig 3(c): Common cause relationship

Fig 3(d): Common effect relationship

- **Indirect causal effect**: Fig 3(a) shows an indirect causal relationship between variables **X** and **Y**. For intuition, let's consider the late-for-school model, where $A \rightarrow J \rightarrow L$ is a causal relationship. Let's first consider the case where J is not observed. If we observe that there has been an accident, then it increases our belief that there would be a traffic jam, which eventually leads to an increase in the probability of getting late for school. Here we see that if the variable J is not observed, then A is able to influence L through J. However, if we consider the case where J is observed, say we have observed that there is a traffic jam, then irrespective of whether there has been an accident or not, it won't change our belief of getting late for school. Therefore, in this case we see that $A \perp L \mid J$.

  More often, in the case of an indirect causal relationship $X \perp Y \mid Z$.

- **Indirect evidential effect**: Fig 3(b) represents an indirect evidential relationship. In the late-for-school model, we can again take the example of $L \rightarrow J \leftarrow A$. Let's first take the case where we haven't observed J. Now, if we observe that somebody is late for school, it increases our belief that there might be a traffic jam, which increases our belief about there being an accident. This leads us to the same results as we got in the case of an indirect causal effect. The variables X and Y are dependent, but become independent if we observe Z, that is $X \perp Y \mid Z$.

- **Common cause**: Fig 3(c) represents a common cause relationship. Let's take the example of $L \leftarrow J \rightarrow Q$ from our late-for-school model. Taking the case where J is not observed, we see that getting late for school makes our belief of being in a traffic jam stronger, which also leads to an increase in the probability of being in a long queue. However, what if we already have observed that there was a traffic jam? In this case, getting late for school doesn't have any effect on being in a long queue. Hence, we see that the independence conditions in this case are also the same as we saw in the previous two cases, that is, X is able to influence Y through Z only if Z is not observed.

- **Common effect**: Fig 3(d) represents a common effect relationship. Taking the example of $A \rightarrow J \leftarrow B$ from the late-for-school model, if we have an observation that there was an accident, it increases the probability of having a traffic jam, but does not have any effect on the probability of heavy rain. Hence, $A \mid B$. We see that we have a different observation here than the previous three cases. Now, if we consider the case when J is observed, let's say that there has been a jam. If we now observe that there hasn't been an accident, it does increase the probability that there might have been heavy rain. Hence, A is not independent of B if J is observed. More generally, we can say that in the case of common effect, **X** is independent of Y if, and only if, Z is not observed.

Now, in a network, how do we know if a variable influences another variable? Let's say we want to check the independence conditions for X_1 and X_n. Also, let's say they are connected by a trail $X_1 \leftrightarrow X_2 \leftrightarrow \dots \leftrightarrow X_{n-1} \leftrightarrow X_n$ and let Z be the set of observed variables in the Bayesian network. In this case, X_1 will be able to influence X_n if and only if the following two conditions are satisfied:

- For every V structure of the form $X_{i-1} \rightarrow X_i \leftarrow X_{i+1}$ in the trail, either $X_i \epsilon Z$ or any descendant of X_i is an element of Z
- No other node on the trail is in Z

Also, if an influence can flow in a trail in a network, it is known as an *active trail*. Let's see some examples to check the active trails using pgmpy for the late-for-school model:

```
In [28]: model.is_active_trail('accident', 'rain')
Out[28]: False
In [29]: model.is_active_trail('accident', 'rain',
                              observed='traffic_jam')
Out[29]: True
In [30]: model.is_active_trail('getting_up_late', 'rain')
Out[30]: False
In [31]: model.is_active_trail('getting_up_late', 'rain',
                              observed='late_for_school')
Out[31]: True
```

Relating graphs and distributions

In the restaurant example or the late-for-school example, we used the Bayesian network to represent the independencies in the random variables. We also saw that we can use the Bayesian network to represent the joint probability distribution over all the variables using the chain rule. In this section, we will unify these two concepts and show that a probability distribution D can only be represented using a graph G, if and only if D can be represented as a set of CPDs associated with the graph G.

IMAP

A graph object G is called an IMAP of a probability distribution D if the set of independency assertions in G, denoted by $I(G)$, is a subset of the set of independencies in D, denoted by $I(D)$.

Let's take an example of two random variables X and Y with the following two different probability distributions over it:

X	Y	P(X, Y)
x^0	y^0	0.25
x^0	y^1	0.25
x^1	y^0	0.25
x^1	y^1	0.25

In this distribution, we can see that $P(X) = 0.5$ and $P(Y) = 0.5$. Also, $P(X, Y) = P(X)$ $P(Y)$. Hence, the two random variables X and Y are independent. If we try to represent any two random variables using a network, we have three possibilities:

- A graph with two disconnected nodes X and Y
- A graph with an edge from $X \rightarrow Y$
- A graph with an edge from $Y \rightarrow X$

We can see from the previous distribution that $I(D) = \{X \perp Y\}$. In the case of disconnected nodes, we also have $I(G) = \{X \perp Y\}$, whereas for the other two graphs, we have $I(G) = \phi$. Hence, all the three graphs are IMAPS of the distribution, and any of these can be used to represent the probability distribution. However, the graph with both nodes disconnected is able to best represent the probability distribution and is known as the **Perfect Map**.

IMAP to factorization

The structure of the Bayesian network encodes the independencies between the random variables, and every probability distribution for which this BN is an IMAP needs to satisfy these independencies. This allows us to represent the joint probability distribution in a very compact form.

Taking the example of the late-for-school model, using the chain rule, we can show that for any distribution, the joint probability distribution would be as follows:

$P(A, R, J, L, S, Q) = P(A) \times P(R|A) \times P(J|A, R) \times P(L|A, R, J) \times P(S|A, R, J, L) \times$

$P(Q|A, R, J, L, S)$

However, if we consider a distribution for which the BN is an IMAP, we get information about the independencies in the distribution. As we can see in this example, we know from the Bayesian network structure that S is independent of A and R, given J and L; Q is independent of A, R, and L, and S, given J; and so on. Applying all these conditions on the equation for joint probability distribution reduces it to the following:

$$P(A, R, J, L, S, Q) = P(A) \times P(R) \times P(J|A, R) \times P(L) \times P(S|J, L) \times P(Q|J)$$

Every graph object has associated independencies with it. These independencies allow us to represent the joint probability distribution of the BN in a compact form.

CPD representations

Till now, we have only been working with tabular CPDs. In a tabular CPD, we take all the possible combinations of different states of a variable and represent them in a tabular form. However, in many cases, tabular CPD is not the best choice to represent CPDs. We can take the example of a continuous random variable. As a continuous variable doesn't have states (or let's say infinite states), we can never create a tabular representation for it. There are many other cases which we will discuss in this section when other types of representation are a better choice.

Deterministic CPDs

One of the cases when the tabular CPD isn't a good choice is when we have a deterministic random variable, whose value depends only on the values of its parents in the model. For such a variable X with parents $Par(X)$, we have the following:

$$P\left(X \mid Par\left(X\right)\right) = \begin{cases} 1 \text{ if } x = f\left(Par\left(X\right)\right) \\ 0 \text{ otherwise} \end{cases}$$

Here, $f : Val\left(Par\left(X\right)\right) \rightarrow Val\left(X\right)$.

We can take the example of logic gates (AND, OR, and so on), where the output of the gate is deterministic in nature and depends only on its inputs. We represent it as a Bayesian network, as shown in Fig 1.7:

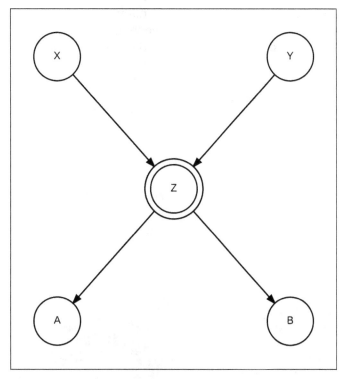

Fig 1.7: A Bayesian network for a logic gate. X and Y are the inputs, A and B are the outputs and Z is a deterministic variable representing the operation of the logic gate.

Here, X and Y are the inputs to the logic gate and Z is the output. We usually denote a deterministic variable by double circles. We can also see that having a deterministic variable gives up more information about the independencies in the network. If we are given the values of X and Y, we know the value of Z, which leads us to the assertion $\rho_1 : \left(T^1, L^0 : 0 \right)$.

Context-specific CPDs

We saw the case of deterministic variables where there was a structure in the CPD, which can help us reduce the size of the whole CPD table. As in the case of deterministic variables, structure may occur in many other problems as well. Think of adding a variable *Flat Tyre* to our late-for-school model. If we have a **Flat Tyre (F)**, irrespective of the values of all other variables, the value of the *Late for school* variable is always going to be 1. If we think of representing this situation using a tabular CPD, we will have all the values for *Late for school* corresponding to $F = 1$ that will be 1, which would essentially be half the table. Hence, if we use tabular CPD, we will be wasting a lot of memory to store values that can simply be represented by a single condition. In such cases, we can use the Tree CPD or Rule CPD.

Tree CPD

A great option to represent such context-specific cases is to use a tree structure to represent the various contexts. In a Tree CPD, each leaf represents the various possible conditional distributions, and the path to the leaf represents the conditions for that distribution. Let's take an example by adding a Flat Tyre variable to our earlier model, as shown in Fig 1.8:

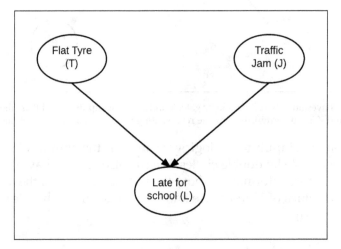

Fig 1.8: Network after adding Flat Tyre (T) variable

If we represent the CPD of L using a Tree CPD, we will get something like this:

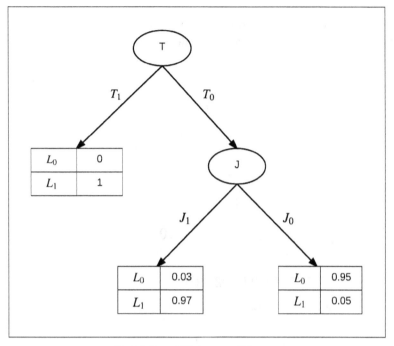

Fig 1.9: Tree CPD in case of Flat tyre

Here, we can see that rather than having four values for the CPD, which we would have to store in the case of Tabular CPD, we only need to store three values in the case of the Tree CPD. This improvement doesn't seem very significant right now, but when we have a large number of variables with high cardinalities, there is a very significant improvement.

Now, let's see how we can implement this using pmgpy:

```
In [1]: from pgmpy.factors import TreeCPD, Factor
In [2]: tree_cpd = TreeCPD([
                ('B', Factor(['A'], [2], [0.8, 0.2]), '0'),
                ('B', 'C', '1'),
                ('C', Factor(['A'], [2], [0.1, 0.9]), '0'),
                ('C', 'D', '1'),
                ('D', Factor(['A'], [2], [0.9, 0.1]), '0'),
                ('D', Factor(['A'], [2], [0.4, 0.6]), '1')])
```

 pgmpy also supports Tree CPDs, where each node has more than one variable.

Rule CPD

Rule CPD is another more explicit form of representation of CPDs. Rule CPD is basically a set of rules along with the corresponding values of the variable. Taking the same example of Flat Tyre, we get the following Rule CPD:

$$\rho_1 : \left(T^1, L^0 : 0\right)$$

$$\rho_2 : \left(T^1, L^1 : 1\right)$$

$$\rho_3 : \left(T^0, J^1, L^0 : 0.95\right)$$

$$\rho_4 : \left(T^1, J^1, L^1 : 0.05\right)$$

$$\rho_5 : \left(T^0, J^0, L^0 : 0.03\right)$$

$$\rho_6 : \left(T^0, J^0, L^1 : 0.97\right)$$

Let's see the code implementation using `pgmpy`:

```
In [1]: from pgmpy.factors import RuleCPD
In [2]: rule = RuleCPD('A', {('A_0', 'B_0'): 0.8,
                             ('A_1', 'B_0'): 0.2,
                             ('A_0', 'B_1', 'C_0'): 0.4,
                             ('A_1', 'B_1', 'C_0'): 0.6,
                             ('A_0', 'B_1', 'C_1'): 0.9,
                             ('A_1', 'B_1', 'C_1'): 0.1})
```

Summary

In this chapter, we saw how we can represent a complex joint probability distribution using a directed graph and a conditional probability distribution associated with each node, which is collectively known as a Bayesian network. We discussed the various reasoning patterns, namely causal, evidential, and intercausal, in a Bayesian network and how changing the CPD of a variable affects other variables. We also discussed the concept of IMAPS, which helped us understand when a joint probability distribution can be encoded in a graph structure.

In the next chapter, we will see that when the relationship between the variables are not causal, a Bayesian model is not sufficient to model our problems. To work with such problems, we will introduce another type of undirected model, known as a Markov model.

2

Markov Network Fundamentals

In the previous chapter, we saw how we can represent a **joint probability distribution** (**JPD**) using a directed graph and a set of **conditional probability distributions** (**CPDs**). However, it's not always possible to capture the independencies of a distribution using a Bayesian model. In this chapter, we will introduce undirected models, also known as Markov networks. We generally use Markov networks when we can't naturally define directionality in the interaction between random variables.

In this chapter, we will cover:

- The basics of factors and their operations
- The Markov model and Gibbs distribution
- The factor graph
- Independencies in the Markov model
- Conversion of the Bayesian model to the Markov model and vice versa
- Chordal graphs and triangulation heuristics

Introducing the Markov network

Let's take an example of four people who go out for dinner in different groups of two. *A* goes out with *B*, *B* goes out with *C*, *C* with *D*, and *D* with *A*. Due to some reasons (maybe due to a bad relationship), *B* doesn't want to go with *D*, and the same holds true for *A* and *C*. Let's think about the probability of them ordering food of the same cuisine. From our social experience, we know that people interacting with each other may influence each other's choice of food. In general, we can say that if *A* influences *B*'s choice and *B* influences *C*'s, then *A* might (as it is probabilistic) indirectly be influencing *C*'s choice. However, given *B*'s and *D*'s choices, we can say with confidence that *A* won't affect *C*'s choice of food. Formally, we can put this as $A \perp C | B, D$. Similarly, $B \perp D | A, C$ as there is no direct interaction between *A* and *C* nor between *B* and *D*.

Let's try to model these independencies using a Bayesian network:

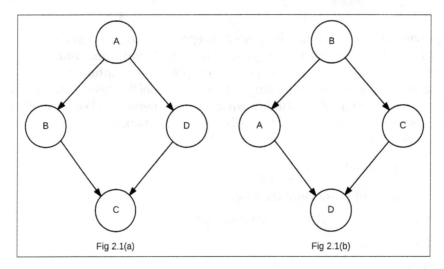

Fig 2.1(a)   Fig 2.1(b)

In the preceding figure, the one labeled Fig 2.1(a) is the Bayesian network representing $A \perp C | B, D$, whereas the one labeled Fig 2.1(b) is the Bayesian network representing $B \perp D | A, C$

The first Bayesian network, Fig 2.1(a), satisfied the first independence assertion, that is, $A \perp C | B, D$, but it couldn't satisfy the second one. Similarly, the second Bayesian network, Fig 2.1(b), satisfied the independence assertion $B \perp D | A, C$, but not the other one. Thus, neither of them is an I-Map for the distribution. Hence, we see that directed models have a limitation and there are conditions that we are unable to represent using directed models.

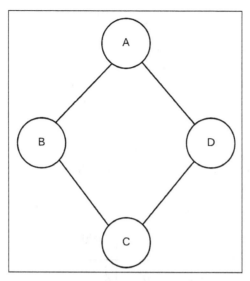

Fig 2.2 Undirected graphical model encoding independencies $A \perp C \mid B, D$ and $B \perp D \mid A, C$

To correctly represent these independencies, we require an undirected model, also known as a Markov network. These are similar to the Bayesian network, in the sense that we represent all the random variables in the form of nodes, but we represent the dependencies or interaction between these random variables with an undirected edge. Before we go into the representation of these models, we need to think about the parameterization of these models. In the Bayesian network, we had a CPD $P(X_i \mid Par(X_i))$ associated with every node X_i. As we don't have any directional influence or a parent-children relationship in the case of the Markov network, instead of using CPDs, we use a more symmetric representation called **factor**, which basically represents how likely it is for some states of a variable to agree with the states of other variables.

Parameterizing a Markov network – factor

Formally, a factor ϕ of a set of random variables X is defined as a function mapping values of X to some real number $\mathbb{R}$:

$$\phi(X) : Val(X) \rightarrow \mathbb{R}$$

Unlike CPDs, there is no notion of directionality or causal relationship among random variables in factors. Factors help in symmetric parameterization of random variables. As the values in a factor don't represent the probability, they are not constrained to sum up to 1 or to be in the range [0,1]. In general, they represent the similarities (or, sometimes, compatibility) among the random variables. Therefore, the higher the value of a combination of states, the greater the compatibility for those states of variables. For example, if we say that two binary random variables A and B are likely to be in the same state rather than different states, we can have a factor where $\phi(a^0, b^0) > \phi(a^0, b^1)$, $\phi(a^0, b^0) > \phi(a^1, b^0)$, $\phi(a^1, b^1) > \phi(a^1, b^0)$, and $\phi(a^1, b^1) > \phi(a^0, b^1)$. This situation can be represented by a factor as follows:

A	B	$\phi(A, B)$
a^0	b^0	1000
a^0	b^1	1
a^1	b^0	5
a^1	b^1	100

We also define the scope of a factor to be the set of random variables over which it is defined. For example, the scope of the preceding factor is {A, B}.

In pgmpy, we can define factors in the following way:

```
# Firstly we need to import Factor
In [1]: from pgmpy.factors import Factor

# Each factor is represented by its scope,
# cardinality of each variable in the scope and their values
In [2]: phi = Factor(['A', 'B'], [2, 2], [1000, 1, 5, 100])
```

In pgmpy, the order in which variables are passed to the factor also has significance. The entries in the factors assume that the random variables on the right change more frequently than the ones present on left (as represented in the previous example).

Now let's try printing a factor:

```
In [3]: print(phi)
```

a	b	phi(A,B)
A_0	B_0	1000.0000
A_0	B_1	1.0000
A_1	B_0	5.0000
A_1	B_1	100.0000

Factors subsume the notion of CPD. So, in pgmpy, CPD classes such as TabularCPD, TreeCPD, and RuleCPD are derived from the Factor class.

Factor operations

There are numerous mathematical operations on factors; the major ones are marginalization, reduction, and product.

The marginalization of a factor is similar to its probabilistic marginalization. If we marginalize a factor ϕ whose scope is W with respect to a set of random variables X, it means to sum out all the entries of X, to reduce its scope to $\{W - X\}$. Here's an example for marginalizing a factor:

```
# In the preceding example phi, let's try to marginalize it with
# respect to B
In [4]: phi_marginalized = phi.marginalize('B', inplace=False)
In [5]: phi_marginalized.scope()
Out[6]: ['A']
# If inplace=True (default), it would modify the original factor
# instead of returning a new one
In [7]: phi.marginalize('A')
In [8]: print(phi)
```

```
+-----+-----------+
| B   |   phi(B)  |
+=====+===========+
| B_0 | 1005.0000 |
+-----+-----------+
| B_1 |  101.0000 |
+-----+-----------+
```

```
In [9]: phi.scope()
Out[9]: ['B']
```

```
# A factor can be also marginalized with respect to more than one
# random variable
```

```
In [10]: price = Factor(['price', 'quality', 'location'],
                        [2, 2, 2],
                        np.arange(8))
In [11]: price_marginalized = price.marginalize(
                                ['quality', 'location'],
                                inplace=False)
In [12]: price_marginalized.scope()
Out[12]: ['price']
In [13]: print(price_marginalized)
```

price	phi(price)
price_0	6.0000
price_1	22.0000

Reduction of a factor ϕ whose scope is W to the context x^i means removing all the entries from the factor where $X = x^i$. This reduces the scope to $W - X$, as ϕ no longer depends on X.

```
# In the preceding example phi, let's try to reduce to the context
# of b_0
In [14]: phi = Factor(['a', 'b'], [2, 2], [1000, 1, 5, 100])
In [15]: phi_reduced = phi.reduce(('b', 0), inplace=False)
In [16]: print(phi_reduced)
```

a	phi(a)
a_0	1000.0000
a_1	5.0000

```
In [17]: phi_reduced.scope()
Out[17]: ['a']
```

```
# If inplace=True (default), it would modify the original factor
# instead of returning a new object.
In [18]: phi.reduce(('a', 1))
In [19]: print(phi)
```

b	phi(b)
b_0	5.0000
b_1	100.0000

```
In [20]: phi.scope()
Out[20]: ['b']

# A factor can be also reduced with respect to more than one
# random variable
In [21]: price_reduced = price.reduce(
                          [('quality', 0), ('location', 1)],
                          inplace=False)
In [22]: price_reduced.scope()
Out[22]: ['price']
```

The term factor product refers to the product of factors ϕ_1 with a scope X and ϕ_2 with scope Y to produce a factor ϕ with a scope $X \cup Y$:

```
In [23]: phi1 = Factor(['a', 'b'], [2, 2], [1000, 1, 5, 100])
In [24]: phi2 = Factor(['b', 'c'], [2, 3],
                          [1, 100, 5, 200, 3, 1000])
# Factors product can be accomplished with the * (product)
# operator
In [25]: phi = phi1 * phi2
In [26]: phi.scope()
Out[26]: ['a', 'b', 'c']
In [27]: print(phi)
```

a	b	c	phi(a,b,c)
a_0	b_0	c_0	1000.0000
a_0	b_0	c_1	100000.0000
a_0	b_0	c_2	5000.0000
a_0	b_1	c_0	200.0000
a_0	b_1	c_1	3.0000
a_0	b_1	c_2	1000.0000
a_1	b_0	c_0	5.0000
a_1	b_0	c_1	500.0000
a_1	b_0	c_2	25.0000
a_1	b_1	c_0	20000.0000
a_1	b_1	c_1	300.0000
a_1	b_1	c_2	100000.0000

```
# or with product method
In [28]: phi_new = phi.product(phi1, phi2)
# would produce a factor with phi_new = phi * phi1 * phi2
```

Gibbs distributions and Markov networks

In the previous section, we saw that we use factors to parameterize a Markov network, which is quite similar to CPDs. Hence, we may think that factors behave in the same way as the CPD. Marginalizing and normalizing it may represent the probability of a variable, but this intuition turns out to be wrong, as we will see in this section. A single factor is just one contribution to the overall joint probability distribution; to have a joint distribution over all the variables, we need the contributions from all the factors of the model. For the dinner example, let's consider the following factors to parameterize the network.

Factor over the variables A and B represented by $\phi(A,B)$:

A	B	$\phi(A,B)$
a^0	b^0	90
a^0	b^1	100
a^1	b^0	1
a^1	b^1	10

Factor over the variables B and C represented by $\phi(B,C)$:

B	C	$\phi(B,C)$
b^0	c^0	10
b^0	c^1	80
b^1	c^0	70
b^1	c^1	30

Factor over variables C and D represented by $\phi(C,D)$:

C	D	$\phi(C,D)$
c^0	d^0	10
c^0	d^1	1
c^1	d^0	100
c^1	d^1	90

Factor over variables A and D represented by $\phi(D,A)$:

D	A	$\phi(D,A)$
d^0	a^0	80
d^0	a^1	60
d^1	a^0	20
d^1	a^1	10

Now, considering the preceding factors, let's try to calculate the probability of A by just considering the factor $\phi(A,B)$. On normalizing and marginalizing the factor with respect to B, we get $P(a^0) = 0.945$ and $P(a^1) = 0.055$. Now, let's try to compute the probability by considering all the factors. To do this, we first need to calculate the factor product of these factors:

A	B	C	D	$\phi(A,B,C,D)$	$1/Z* \phi(A,B,C,D)$
a^0	b^0	c^0	d^0	$8*10^3$	$1.0360*10-4$
a^0	b^0	c^0	d^1	$0.2*10^3$	$2.5901*10-6$
a^0	b^0	c^1	d^0	$64*10^3$	$8.2883*10-4$
a^0	b^0	c^1	d^1	$16*10^3$	$2.0720*10-4$

A	B	C	D	$\phi(A,B,C,D)$	$1/Z*\phi(A,B,C,D)$
a^0	b^1	c^0	d^0	$56*10^5$	$7.2522*10-2$
a^0	b^1	c^0	d^1	$14*10^4$	$1.8130*10-3$
a^0	b^1	c^1	d^0	$24*10^6$	$3.1081*10-1$
a^0	b^1	c^1	d^1	$6*10^5$	$7.7702*10-3$
a^1	b^0	c^0	d^0	$54*10^4$	$6.9932*10-3$
a^1	b^0	c^0	d^1	$9*10^3$	$1.1655*10-4$
a^1	b^0	c^1	d^0	$432*10^5$	$5.5946*10-1$
a^1	b^0	c^1	d^1	$72*10^4$	$9.3243*10-3$
a^1	b^1	c^0	d^0	$42*10^4$	$5.4392*10-3$
a^1	b^1	c^0	d^1	$7*10^3$	$9.0653*10-4$
a^1	b^1	c^1	d^0	$18*10^5$	$2.3310*10-2$
a^1	b^1	c^1	d^1	$3*10^4$	$3.8851*10-4$

Now, if we normalize and marginalize this factor product, we get $P(a^0)=0.3940$ and $P(a^1)=0.6059$. Here, we see that there is a huge difference in the probability when we consider only a single factor as compared to when we consider all the factors. Hence, our intuition of factors behaving like CPDs is wrong.

Therefore, in a Markov network over a set of variables $X = \{X_1, X_2,..., X_m\}$ having a set of factors $\Phi = \{\phi_1, \phi_2,..., \phi_n\}$ associated with it, we can compute the joint probability distribution over these variables as follows:

$$P(X_1, X_2,..., X_m) = \frac{1}{Z}\prod_{\phi \in \Phi}\phi$$

Here, Z is the partition function and $Z = \sum_{X_1,X_2,...,X_n}\prod_{\phi \in \Phi}\phi$.

Also, a distribution P_ϕ is called a Gibbs distribution parameterized by a set of factors $\Phi = \{\phi_1(D_1), \phi_2(D_2), ..., \phi_n(D_n)\}$ if it is defined as follows:

$$P_\phi(X_1, X_2, X_3, ..., X_n) = \frac{1}{Z}\left(\phi_1(D_1) * \phi_2(D_2) * \phi_3(D_3) * ... * \phi_n(D_n)\right)$$

Here, $Z = \sum_{X_1, X_2, ..., X_n}\left(\phi_1(D_1) * \phi_2(D_2) * \phi_3(D_3) * ... * \phi_n(D_n)\right)$ is a normalizing constant called the partition function.

To construct a Markov network, we need to associate the parameterization of a Gibbs distribution to the set of factors of an undirected graph structure. A factor with the X and Y scopes represents a direct relationship between them.

Let's see how we can represent a Markov model using `pgmpy`:

```
# First import MarkovModel class from pgmpy.models
In [1]: from pgmpy.models import MarkovModel
In [2]: model = MarkovModel([('A', 'B'), ('B', 'C')])
In [3]: model.add_node('D')
In [4]: model.add_edges_from([('C', 'D'), ('D', 'A')])
```

Now, let's try to define a few factors to associate with this model:

```
In [5]: from pgmpy.factors import Factor
In [6]: factor_a_b = Factor(variables=['A', 'B'],
                            cardinality=[2, 2],
                            value=[90, 100, 1, 10])
In [7]: factor_b_c = Factor(variables=['B', 'C'],
                            cardinality=[2, 2],
                            value=[10, 80, 70, 30])
In [8]: factor_c_d = Factor(variables=['C', 'D'],
                            cardinality=[2, 2],
                            value=[10, 1, 100, 90])
In [9]: factor_d_a = Factor(variables=['D', 'A'],
                            cardinality=[2, 2],
                            value=[80, 60, 20, 10])
```

We can associate the factors to the model using the `add_factors` method:

```
In [10]: model.add_factors(factor_a_b, factor_b_c,
                           factor_c_d, factor_d_a)
In [11]: model.get_factors()
Out[11]:
[<Factor representing phi(A:2, B:2) at 0x7f18504477b8>,
 <Factor representing phi(B:2, C:2) at 0x7f18504479b0>,
 <Factor representing phi(C:2, D:2) at 0x7f1850447f98>,
 <Factor representing phi(D:2, A:2) at 0x7f1850455358>]
```

The factor graph

The Markov network doesn't give a very clear picture of the Gibbs parameterization of the distribution because we can't conclude whether the factors in it involve the maximal cliques or subgraphs. To overcome this limitation of the Markov network, we require a representation that can show the parameterization explicitly. The factor graph is one such representation.

A factor graph is a bipartite graph, one disjoint set being variable nodes, representing the variables, and the other being factor nodes, representing factors. An edge between a variable node and a factor node denotes that the random variable belongs to the scope of the factor. Thus, a factor graph is parameterized by a set of factors, where each of them is associated with a factor node, whose scope is all sets of all the random variables that it is neighbor to.

Generally, all the variable nodes are represented by a circle and all the factor nodes are represented by a square. Here's an example:

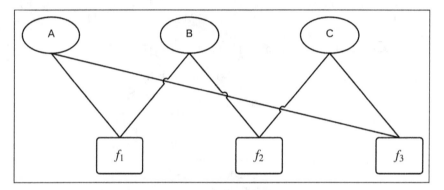

Fig 2.3 Factor graph

In the preceding factor graph, there are three variable nodes A, B, and C and three factor nodes associated with three factors, namely $\phi_1(A,B)$, $\phi_2(B,C)$, and $\phi_3(C,A)$. This representation is more explicit than the Markov network (Fig 2.4 (a)). From the Markov network, without checking the factors, we can't know whether the factors involve maximal clique *(A, B, C)* or their subgraphs *{(A, B), (B, C), (C, A)}*. This information is explicitly specified in the factor graph.

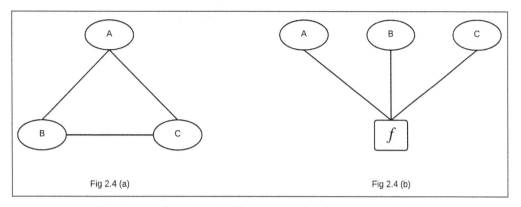

Fig 2.4 (a)   Fig 2.4 (b)

Fig 2.4 (a) Markov network of the corresponding factor graph in Fig 2.3

Fig 2.4 (b) Factor graph parameterized by factors involving maximal clique of the Markov network

In pgmpy, factor graphs can be created as follows:

```
# First import FactorGraph class from pgmpy.models
In [1]: from pgmpy.models import FactorGraph
In [2]: factor_graph = FactorGraph()

# Add nodes (both variable nodes and factor nodes) to the model
# as we did in previous other models
In [3]: factor_graph.add_nodes_from(['A', 'B', 'C', 'D',
                                    'phi1', 'phi2', 'phi3'])

# Add edges between all variable nodes and factor nodes
In [4]: factor_graph.add_edges_from(
                    [('A', 'phi1'), ('B', 'phi1'),
                     ('B', 'phi2'), ('C', 'phi2'),
                     ('C', 'phi3'), ('A', 'phi3')])
```

The FactorGraph class doesn't make any prior assumption about nodes; that is, it doesn't know which nodes are variable nodes and which nodes are factor nodes. It allows us to add edges between any nodes as long as they don't violate the bipartite nature of the factor graph. As soon as the bipartite property is violated by the addition of an edge, it raises the ValueError exception:

```
# We can also add factors into the model
In [5]: from pgmpy.factors import Factor
In [6]: import numpy as np
In [7]: phi1 = Factor(['A', 'B'], [2, 2], np.random.rand(4))
In [8]: phi2 = Factor(['B', 'C'], [2, 2], np.random.rand(4))
In [9]: phi3 = Factor(['C', 'A'], [2, 2], np.random.rand(4))
In [10]: factor_graph.add_factors(phi1, phi2, phi3)
```

We can also convert a Markov model into a factor graph and vice versa:

```
In [11]: from pgmpy.models import MarkovModel
In [12]: mm = MarkovModel()
In [13]: mm.add_nodes_from(['A', 'B', 'C'])
In [14]: mm.add_edges_from([('A', 'B'), ('B', 'C'), ('C', 'A')])
In [15]: mm.add_factors(phi1, phi2, phi3)
In [16]: factor_graph_from_mm = mm.to_factor_graph()

# While converting a markov model into factor graph, factor nodes
# would be automatically added the factor nodes would be in the
# form of phi_node1_node2_...

In [17]: factor_graph_from_mm.nodes()
Out[17]: ['C', 'B', 'phi_A_B', 'phi_B_C', 'phi_C_A', 'C']
In [18]: factor_graph.edges()
Out[18]: [('phi_A_B', 'A'), ('phi_A_C', 'A'), ('B', 'phi_B_C'),
          ('B', 'phi_A_B'), ('C', 'phi_B_C'), ('C', 'phi_C_A')]

# FactorGraph to MarkovModel

In [19]: phi = Factor(['A', 'B', 'C'], [2, 2, 2],

np.random.rand(8))
In [20]: factor_graph = FactorGraph()
In [21]: factor_graph.add_nodes_from(['A', 'B', 'C', 'phi'])
In [22]: factor_graph.add_edges_from(

                 [('A', 'phi'), ('B', 'phi'), ('C', 'phi')])
In [23]: mm_from_factor_graph = factor_graph.to_markov_model()
In [24]: mm_from_factor_graph.add_factors(phi)
In [24]: mm_from_factor_graph.edges()
Out[24]: [('B', 'A'), ('C', 'B'), ('C', 'A')]
```

Independencies in Markov networks

In the previous chapter, we saw how a Bayesian network structure encodes independency conditions in it, and how observing variables affects the flow of influence in the network. Similarly, in the case of Markov networks, the graph structure encodes independency conditions. However, the flow of influence in a Markov network stops as soon as any node is observed in that trail. This is quite different from what we saw in the Bayesian network, where different structures responded differently to the observation of the nodes.

To understand this more formally, let H be a Markov network structure and $Z \subseteq X$ be a set of observed variables. Then, the path $X_1 X_2 ... X_{k-1} X_k$ is active if and only if none of the X_i for $i \epsilon \{1, 2, ..., k-1, k\}$ are in Z.

In the case of Bayesian networks, we had the concept of local independencies, where a variable is independent of all its non-descendants, given given its parents. We also had global conditions which were implied by **D-Separation**. Similarly, in the case of Markov networks, the independence conditions that we discussed earlier are the global independencies in the network. Local independence conditions are a subset of global conditions, but local independencies are also very important because they allow us to focus on a much smaller part of the network.

There are two ways of looking at the local independencies in the case of a Markov network. One way is to be intuitive and think that if two nodes X and Y are directly connected, then there is no way of rendering them as independent. However, if they are not directly connected, there is always a way of rendering them conditionally independent of each other. One way to do this is by observing all the variables in the network, except for X and Y. If we have all the nodes observed in the network except X and Y, then there must be at least one observed node in the trail connecting the nodes X and Y, which will eventually lead X and Y to be independent of each other. This is known as **pairwise independency**. More formally, we can define pairwise independency in a Markov network H as follows:

$$I_p(H) = \left\{ \left(X \perp Y \mid \chi - \{X, Y\} \right) \right\}$$

Another way of thinking about local independencies is to not let other nodes influence a given node, by observing all of its neighboring nodes. This set of neighboring nodes is known as the **Markov blanket**, and this type of independence in the network is known as local independency. More formally, this can be defined as follows:

$$I_l(H) = \left\{ \left(X \perp \chi - X - MB_H(X) \mid MB_H(X) \right) \right\}$$

Like Bayesian networks, we also have the concept of I-Map in Markov models. For a probability distribution P and a Markov network structure H if $I(H) \subseteq I(P)$, we say that H is an I-Map of P.

Let's check the local independencies in the network using pgmpy:

```
In [1]: from pgmpy.models import MarkovModel
In [2]: mm = MarkovModel()
In [3]: mm.add_nodes_from(

                 ['x1', 'x2', 'x3', 'x4', 'x5', 'x6', 'x7'])
In [4]: mm.add_edges_from(

                 [('x1', 'x3'), ('x1', 'x4'), ('x2', 'x4'),

                  ('x2', 'x5'), ('x3', 'x6'), ('x4', 'x6'),
                  ('x4', 'x7'), ('x5', 'x7')]])
In [5]: mm.get_local_independencies()
Out[5]:
(x3 _|_ x5, x4, x7, x2 | x6, x1)
(x4 _|_ x3, x5 | x6, x7, x1, x2)
(x1 _|_ x6, x7, x5, x2 | x3, x4)
(x5 _|_ x3, x4, x6, x1 | x7, x2)
(x7 _|_ x3, x6, x1, x2 | x5, x4)
(x2 _|_ x3, x6, x7, x1 | x5, x4)
(x6 _|_ x5, x7, x1, x2 | x3, x4)
```

We saw three different ways of defining independencies in Markov networks. While all of these are related, they are equivalent only for positive distributions. Non-positive distributions allow for deterministic dependencies between the variables, and such deterministic interactions can allow us to construct networks that are not I-maps of the distribution but local independencies hold for them.

Constructing graphs from distributions

To construct a Markov network from a distribution, the mere concept of I-Maps is not enough. As in the case of Bayesian networks, a fully connected graph has no independence conditions and, hence, it can be an I-Map of any probability distribution. Therefore, we introduce the concept of the minimal I-Map in Markov networks as well. To construct a minimal I-Map, we can use the local independency conditions that we defined in the previous section.

In the first approach, let's consider the case of pairwise independencies. According to pairwise independencies, if there is no edge between {X, Y}, then $(X \perp Y \mid \chi - \{X,Y\})$. Thus, at the very least, to guarantee that H is an I-map, we must add direct edges between all pairs of nodes X and Y, such that they are dependent even on observing all the other variables in the network.

Similarly, we can get more information about the structure by using the local independencies conditions. For each variable X, we can find the minimal set of nodes. Observing these makes the variable independent of all the variables. Then, add an edge between the variable and all the nodes in the set. In this way, exploiting local independencies gives us a very basic methodology for constructing models from data. In later chapters, we will discuss more sophisticated methods to create models.

Bayesian and Markov networks

Until now, we have discussed two different models for representing graphical models. Each of these can represent independence constraints that the other cannot. In this section, we will look at the relationship between these two models.

Converting Bayesian models into Markov models

Both Bayesian models and Markov models parameterize a probability distribution using a graphical model. Further, these structures also encode the independencies among the random variable. So, when converting a Bayesian model into a Markov one, we have to look from the following two perspectives:

- From the perspective of parameterization, that is, representing the probability distribution of the Bayesian model P_B using a fully parameterized Markov model

- From the perspective of independencies, that is, representing the independence constraints encoded by the Bayesian model using the Markov model

From the first perspective, conversion of the Bayesian model into the Markov model is fairly simple. Let's begin by considering the case of a probability distribution P_B, where B is a parameterized Bayesian network over a graph G. If we see the parameterization of the Bayesian network, we can also think of it as a parameterization of a Gibbs distribution. We can think of a CPD over a variable X_i to be a factor with a scope $\{X_i, Pa_{Xi}\}$. This set of factors defines a Gibbs distribution with the partition function being equal to 1.

Looking from the second perspective, let's try to find out what kind of undirected graph would be an I-Map for this Gibbs distribution. To understand it more clearly, let's go back to our previous Bayesian network example and try to convert it into a Markov network:

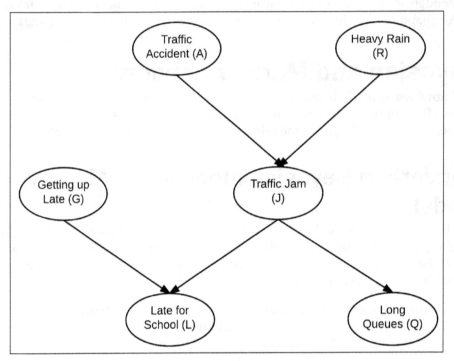

Fig 2.5 Simple Bayesian model

Let's try to convert this Bayesian model into a Markov model simply by replacing directed edges with undirected ones and start by replacing the edges *(A, J)* and *(R, J)* with undirected edges. However, this representation has a problem. The Markov Blanket of node *A* would be *J*. Thus, this representation asserts that *A* would be independent of all the nodes in the model expect *J*, given *J* or specifically $A \perp R \mid J$. However, the Bayesian Network asserts the exact opposite of this, that is, $A \not\perp R \mid J$. Thus, it requires an additional undirected edge between *A* and *R*. Similarly, replacing directed edges with undirected edges and adding extra edges where required, we get the network in the following figure:

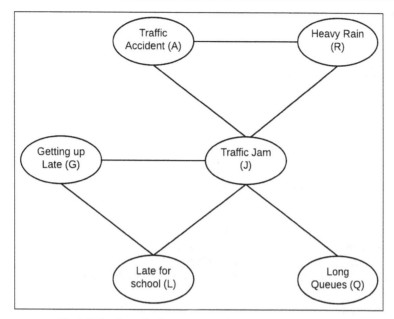

Fig 2.6 Moral graph of Bayesian model represented in Fig 2.5

Hence, we can conclude that to convert a Bayesian model into a Markov model, we need to do the following:

- Replace all the directed edges between the nodes with undirected edges
- Add additional undirected edges between nodes that are parents of the node

This new structure created by replacing directed edges and adding new edges is known as the **moral graph** of the Bayesian network and is also known as the **moralization of the network.**

We can see that the moral graph H of a Bayesian model G loses some information regarding the independencies. For example, $A \perp R$ in the graph G, but not in H. However, $I(H) \subseteq I(G)$, so we can say that H is an I-Map for this Gibbs distribution. Note that moral graphs don't always lose information about the independencies. For example, if there had been an edge between A and R already, then no information regarding independencies would have been lost.

In pgmpy, a Bayesian model can be converted into a Markov model as follows:

```
In [1]: from pgmpy.models import BayesianModel
In [2]: from pgmpy.factors import TabularCPD

# Creating the above bayesian network
In [2]: model = BayesianModel()
In [3]: model.add_nodes_from(['Rain', 'TrafficJam'])
```

```
In [4]: model.add_edge('Rain', 'TrafficJam')
In [5]: model.add_edge('Accident', 'TrafficJam')
In [6]: cpd_rain = TabularCPD('Rain', 2, [[0.4], [0.6]])
In [7]: cpd_accident = TabularCPD('Accident', 2, [[0.2], [0.8]])
In [8]: cpd_traffic_jam = TabularCPD(

                'TrafficJam', 2,

                [[0.9, 0.6, 0.7, 0.1],

                 [0.1, 0.4, 0.3, 0.9]],

                evidence=['Rain', 'Accident'],

                evidence_card=[2, 2])
In [9]: model.add_cpds(cpd_rain, cpd_accident, cpd_traffic_jam)
In [10]: model.add_node('LongQueues')
In [11]: model.add_edge('TrafficJam', 'LongQueues')
In [12]: cpd_long_queues = TabularCPD('LongQueues', 2,

                                        [[0.9, 0.2],

                                         [0.1, 0.8]],
                                        evidence=['TrafficJam'],

                                        evidence_card=[2])
In [13]: model.add_cpds(cpd_long_queues)
In [14]: model.add_nodes_from(['GettingUpLate', 'LateForSchool'])
In [15]: model.add_edges_from([('GettingUpLate', 'LateForSchool'),

                ('TrafficJam', 'LateForSchool')])
In [16]: cpd_getting_up_late = TabularCPD('GettingUpLate', 2,

                                        [[0.6], [0.4]])
In [17]: cpd_late_for_school = TabularCPD(

                'LateForSchool', 2,

                [[0.9, 0.45, 0.8, 0.1],

                 [0.1, 0.55, 0.2, 0.9]],

                evidence=['GettingUpLate',TrafficJam'],

                evidence_card=[2, 2])
In [18]: model.add_cpds(cpd_getting_up_late, cpd_late_for_school)

# Conversion from BayesianModel to MarkovModel is accomplished by

In [19]: mm = model.to_markov_model()
```

```
In [20]: mm.edges()
Out[20]:
[('TrafficJam', 'Accident'),
 ('TrafficJam', 'LongQueues'),
 ('TrafficJam', 'LateForSchool'),
 ('TrafficJam', 'Rain'),
 ('TrafficJam', 'GettingUpLate'),
 ('LateForSchool', 'GettingUpLate'),
 ('Accident', 'Rain')]
```

Converting Markov models into Bayesian models

The conversion of a Markov model into a Bayesian model is not as simple as the case of converting a Bayesian model into a Markov model.

Let's start with our simple Markov model example and try to convert it into a Bayesian model. In this section, we will be looking from the perspective of independencies, that is, finding a Bayesian model that is an I-Map of the corresponding Markov model:

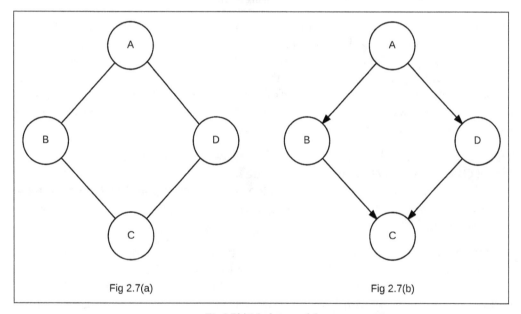

Fig 2.7(a) Markov model

Fig 2.7(b) Bayesian model formed by changing the directed edges into undirected ones

We can simply replace all the undirected edges in the Markov model (Fig 2.7(a)) with directed edges (Fig 2.7(b)). However, does this Bayesian model encode all the independencies of the corresponding Markov model? Before getting into this, let's take a more simple example of the Markov model formed by removing the node C:

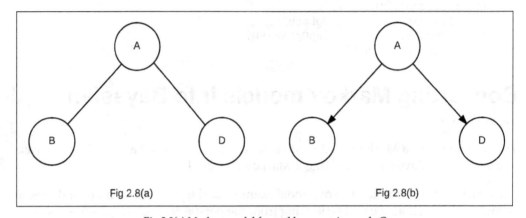

Fig 2.8(a)   Fig 2.8(b)

Fig 2.8(a) Markov model formed by removing node C

Fig 2.8(b) Bayesian model formed by changing the directed edges into undirected ones

Fig 2.8(a) represents a Markov model formed by removing the node C. Fig 2.8(b) is formed just by converting the undirected edges into directed edges. The Markov model encodes the independence assertion that $B \perp D \mid A$, which is also encoded in the corresponding Bayesian model. So, the Bayesian model formed is a perfect I-Map of the Markov model. Now, let's go back to our previous example and examine the independencies encoded in both, the Markov model and the Bayesian model formed simply by converting undirected edges into directed ones.

The Markov model H encodes $B \perp D \mid A, C$, but the corresponding Bayesian model G encodes $B \perp D \mid A$, which is not true for H, where $B \not\perp D \mid A$. So, for G to be an I-Map for H, there should be a directed edge between B and D. However, why did simply converting the undirected edges into direct edges not suffice as in the example in Fig 2.8?

We can see that the example in Fig 2.7 is a non-triangulated (non-chordal) graph. A triangulated or chordal graph is a graph in which each of its cycles of four or more vertices has a chord (an edge that is not part of the cycle but connects two vertices of the cycle). By simply converting edges of a non-triangulated graph into directed ones, we introduce immoralities. An immorality is a v-structure $(X \rightarrow Z \leftarrow Y)$, if there is no directed edge between X and Y. So why does the introduction of immorality pose an issue? To get the answer to this question, let's look at the example again. Before the introduction of immorality or conversion of edges into directed ones, we had $B \not\perp D \mid A$, but after the addition of immorality, we had $B \perp D \mid A$.

So, we can conclude that the process of converting a Markov models to a Bayesian model requires us to add edges to the network to make it chordal. This process is known as **triangulation**.

In pgmpy, we can convert a Markov model into a Bayesian model in the following way:

```
In [1]: from pgmpy.models import MarkovModel
In [2]: from pgmpy.factors import Factor
In [3]: model = MarkovModel()

# Fig 2.7(a) represents the Markov model
In [4]: model.add_nodes_from(['A', 'B', 'C', 'D'])
In [5]: model.add_edges_from([('A', 'B'), ('B', 'C'),
                              ('C', 'D'), ('D', 'A')])

# Adding some factors.
In [6]: phi_A_B = Factor(['A', 'B'], [2, 2], [1, 100, 100, 1])
In [7]: phi_B_C = Factor(['B', 'C'], [2, 2], [100, 1, 1, 100])
In [8]: phi_C_D = Factor(['C', 'D'], [2, 2], [1, 100, 100, 1])
In [9]: phi_D_A = Factor(['D', 'A'], [2, 2], [100, 1, 1, 100])
In [10]: model.add_factors(phi_A_B, phi_B_C, phi_C_D, phi_D_A)
In [11]: bayesian_model = model.to_bayesian_model()
In [12]: bayesian_model.edges()
Out[12]: [('D', 'C'), ('D', 'B'), ('D', 'A'),
          ('B', 'C'), ('B', 'A')]
```

Chordal graphs

As we have seen, in the case of converting a Bayesian model into a Markov model, we lost some of the independence conditions. The same holds true in this case as well, and we can see from the example that we lose the following conditions:

- Statistical independence between parents of the same node in a Bayesian network is lost when it is converted into a Markov one due to the introduction of immorality

- Addition of extra edges to convert a Markov model into a Bayesian one leads to the loss of local independence information

We also see that for the perfect conversion of the model, we must have a chordal graph. The process of converting a non-chordal graph into a chordal one is called triangulation. A triangulated graph can be obtained from an undirected graph by adding links.

In pgmpy, we can triangulate a graph as follows:

```
In [1]: from pgmpy.models import MarkovModel
In [2]: from pgmpy.factors import Factor
In [3]: import numpy as np
In [4]: model = MarkovModel()

# Fig 2.7(a) represents the MarkovModel
In [6]: model.add_nodes_from(['A', 'B', 'C', 'D'])
In [7]: model.add_edges_from(
            [('A', 'B'), ('B', 'C'), ('C', 'D'), ('D', 'A')])

# Adding some factors
In [8]: phi_A_B = Factor(['A', 'B'], [2, 2], [1, 100, 100, 1])
In [9]: phi_B_C = Factor(['B', 'C'], [2, 2], [100, 1, 1, 100])
In [10]: phi_C_D = Factor(['C', 'D'], [2, 2], [1, 100, 100, 1])
In [11]: phi_D_A = Factor(['D', 'A'], [2, 2], [100, 1, 1, 100])
In [12]: model.add_factors(phi_A_B, phi_B_C, phi_C_D, phi_D_A)
In [13]: chordal_graph = model.triangulate()

# Fig 2.9 represents the chordal graph created by triangulation
In [14]: chordal_graph.edges()
Out[14]: [('C', 'D'), ('C', 'B'), ('D', 'B'),
          ('D', 'A'), ('A', 'B')]
```

The following is the chordal graph formed by triangulating the Markov model:

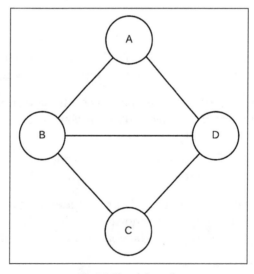

Fig 2.9 Chordal graph

There are six heuristics presented in *Heuristic Algorithms for the Triangulation of Graphs* by Andres Cano and Serafn Moral to add links in an undirected graph to triangulate it. The detailed explanation of these heuristics is beyond the scope of this book (for a detailed explanation, you can go through this paper). These heuristics are also implemented in pgmpy in the following way:

```
# For creating a chordal graph using first heuristic there are six
# heuristics that are implemented in pgmpy and can be used by
# passing the keyword argument heuristic as H1, H2, H3, H4, H5, H6
In [15]: chordal_graph = model.triangulate(heuristic='H1')
```

> If no heuristics are provided, H6 must be used by default.

Summary

In this chapter, we saw how we are not able to use a Bayesian model to model a problem in some cases. In some of these problems, we can use an undirected graph to represent the relation between the variables. These undirected graphs, along with a set of factors representing interaction between these random variables, are known as Markov networks. We discussed the various independencies encoded by a Markov network: local, pairwise, and global. Also, we saw that in a Markov network, the influence stops flowing as soon as we observe any node in that trail, which is quite different from the case of a Bayesian network, where different network structures imply a different flow of influence. We also discussed the concepts of I-Maps and minimal I-Maps that helped us understand when and how to encode a joint probability distribution in a graph structure. We also discussed the relationship between a Bayesian network and a Markov network.

In these first two chapters, we mainly discussed the representation and various properties of Bayesian and Markov models. In the next chapter, we will discuss how to infer the probability values of the different variables when the model is conditioned over some other variables, which would be much like getting predictions for variables for new data points as we do in normal machine learning techniques.

3
Inference – Asking Questions to Models

In the previous chapters, we looked at the different types of models and how to create models for our problems. We also saw how the probabilities of variables change when we change the probabilities of some other variables. In this chapter, we will be discussing the various algorithms that can be used to compute these changes in the probabilities. We will also see how to use these inference algorithms to predict the values of variables of new data points based on our model, which was trained using our previous data.

In this chapter, we will cover:

- Using inference to answer queries about the model
- Variable elimination
- Understanding the belief propagation algorithm using a clique tree
- MAP inference using variable elimination
- MAP inference using belief propagation
- Comparison between variable elimination and belief propagation

Inference

Inferring from a model is the same as finding the conditional probability distribution over some variables, that is, $P(Y \mid E = e)$, where $Y \subseteq \chi$ and $E \subset \chi$. Also, if we think about predicting values for a new data point, we are basically trying to find the conditional probability of the unknown variable, given the observed values of other variables. These conditional distributions can easily be computed from the joint probability distribution of the variables, by marginalizing and reducing them over variables and states.

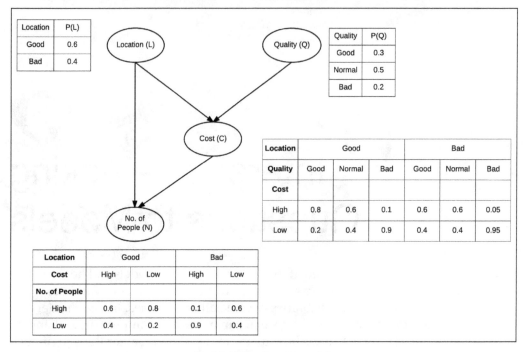

Fig 3.1: The restaurant model

Let's consider the restaurant example once again, as shown in the preceding figure. We can think of various inference queries that we can try on the model. For example, we may want to find the probability of the quality of a restaurant being good, given that the location is good, the cost is high, and the number of people coming is also high, which would result in the probability query $P(Q = good \mid L = good, C = high, N = high)$. Also, if we think of a machine learning problem, where we want to predict the number of people coming to a restaurant given other variables, it would simply be an inference query over the model, and the state having higher probability would be the prediction by the model. Now, let's see how we can compute these conditional probabilities from the model.

From the product rule of probability, we know the following:

$$P(Y \mid E = e) = \frac{P(Y, e)}{P(e)}$$

So, to find each value of the distribution P(Y, e), we could simply do a summation of the joint probability distribution over the variables $W = \chi - Y - E$:

$$P(y,e) = \sum_{w} P(y,e,w)$$

Now, to find P(e), we can simply do another summation over $P(y,e)$, which we just computed:

$$P(e) = \sum_{y} P(y,e)$$

Using these values of $P(y,e)$ and P(e), we can easily find the value of $P(y \mid e)$ as follows:

$$P(y \mid e) = \frac{P(y,e)}{P(e)}$$

Performing a similar calculation for each state y of the random variable Y, we can calculate the conditional distribution over Y, given E = e.

Complexity of inference

In the previous section, we saw how we can find the conditional distributions over variables when a joint distribution is given. However, computing the joint probability distribution will give us an exponentially large table, and avoiding these huge tables was the whole point of introducing probabilistic graphical models. We will be discussing the various algorithms that can help us avoid the complete probability distributions while computing the conditional distribution, but first, let's see what the complexity of computing these inferences is.

If we think about the worst case scenario, we cannot avoid the exponential size of the tables in graphical models, which makes inference an NP-hard problem, and unfortunately, even the approximate methods to compute conditional distributions are NP-hard. Proofs of these results are beyond the scope of this book.

However, these results are for the worst case scenario. In real life, we don't always have the worst case. So, let's discuss various algorithms for the inference.

Variable elimination

Let's try to do some inference tasks over the restaurant network in Fig 3.1. Let's say we want to find *P(C)*. We know the following from the chain rule of probability:

$$P(C) = \sum_{l,q} P(C \mid l,q) P(l,q)$$

Also, we know that the random variables *L* and *Q* are independent of each other if *C* is not observed. So, we can write the preceding equation as follows:

$$P(C) = \sum_{l,q} P(C \mid l,q) P(l) P(q)$$

Now, we can see that we know the probability values involved in the product for the computation of *P(C)*. We have the values of $P(C \mid l,q)$ from the CPD of *C*, the values of *P(l)* from the CPD of *L*, and the values of *P(q)* from the CPD of *Q*. Summing up the product of these probabilities, we can easily find the probability of *C*.

We can also note that the computational cost for this computation would be $\mathcal{O}(Val(C) * Val(L) * Val(Q))$, where $Val(X)$ represents the number of states of the variable *X*. We can see that in order to compute the probability of each state of *C*, we need to compute the product for each combination of states *L* and *Q*, and then add them together. This means that for each state of *C*, we have $2 * Val(L) * Val(P)$ products and $(Val(L) * Val(Q)) - 1$ additions. Here, 2 appears in the number of products because there are two product operations in the equation. Also, we need to do this computation $Val(C)$ times for each state of *C*.

Now, let's take the example of another simple model $A \rightarrow B \rightarrow C \rightarrow D$ and try to find *P(D)*. We can find *P(D)* simply as follows:

$$P(D) \sum_{a,b,c} P(a) P(b \mid a) P(c \mid b) P(D \mid c)$$

However, we can see that the complexity of computing the values of *P(D)* is now $\mathcal{O}(Val(D) * Val(C) * Val(B) * Val(A))$, and for much more complex models, our complexity will be too high. Now, let's see how we can use the concept of dynamic programming to avoid computing the same values multiple times and to reduce our complexity. To see the scope of using dynamic programming in this problem, let's first simply unroll the summation and check what values we are computing. For simplicity, we will assume that each of the variables has only two states. Unrolling the summation, we get the following:

$$P(d^0) = P(a^0)P(b^0 \mid a^0)P(c^0 \mid b^0)P(d^0 \mid c^0) +$$

$$P(a^0)P(b^0 \mid a^0)P(c^1 \mid b^0)P(d^0 \mid c^1) +$$

$$P(a^0)P(b^1 \mid a^0)P(c^0 \mid b^1)P(d^0 \mid c^0) +$$

$$P(a^0)P(b^1 \mid a^0)P(c^1 \mid b^1)P(d^0 \mid c^1) +$$

$$P(a^1)P(b^0 \mid a^1)P(c^0 \mid b^0)P(d^0 \mid c^0) +$$

$$P(a^1)P(b^0 \mid a^1)P(c^1 \mid b^0)P(d^0 \mid c^1) +$$

$$P(a^1)P(b^1 \mid a^1)P(c^0 \mid b^1)P(d^0 \mid c^0) +$$

$$P(a^1)P(b^1 \mid a^1)P(c^1 \mid b^1)P(d^0 \mid c^1)$$

$$P(d^1) = P(a^0)P(b^0 \mid a^0)P(c^0 \mid b^0)P(d^1 \mid c^0) +$$

$$P(a^0)P(b^0 \mid a^0)P(c^1 \mid b^0)P(d^1 \mid c^1) +$$

$$P(a^0)P(b^1 \mid a^0)P(c^0 \mid b^1)P(d^1 \mid c^0) +$$

$$P(a^0)P(b^1 \mid a^0)P(c^1 \mid b^1)P(d^1 \mid c^1) +$$

$$P(a^1)P(b^0 \mid a^1)P(c^0 \mid b^0)P(d^1 \mid c^0) +$$

$$P(a^1)P(b^0 \mid a^1)P(c^1 \mid b^0)P(d^1 \mid c^1) +$$

$$P(a^1)P(b^1 \mid a^1)P(c^0 \mid b^1)P(d^1 \mid c^0) +$$

$$P(a^1)P(b^1 \mid a^1)P(c^1 \mid b^1)P(d^1 \mid c^1)$$

To calculate $P(D)$, we must calculate $P(d^0) =$ and $P(d^1) =$ separately. After unrolling the summations, we can see that we have many computations that we have been doing multiple times if we take the simple linear approach. In the preceding equations, we can see that we have computed $P(a^0) * P(b^0 | a^0)$ four times, $P(a^0) * P(b^1 | a^0)$ four times, and so on. Let's first group these computations together:

$$P(d^0) = P(a^0)P(b^0 | a^0) + P(a^1)P(b^0 | a^1)P(c^0 | b^0)P(d^0 | c^0) +$$

$$P(a^0)P(b^0 | a^0) + P(a^1)P(b^0 | a^1)P(c^1 | b^0)P(d^0 | c^1) +$$

$$P(a^0)P(b^1 | a^0) + P(a^1)P(b^1 | a^1)P(c^0 | b^1)P(d^0 | c^0) +$$

$$P(a^0)P(b^1 | a^0) + P(a^1)P(b^1 | a^1)P(c^1 | b^1)P(d^0 | c^1)$$

$$P(d^1) = P(a^0)P(b^0 | a^0) + P(a^1)P(b^0 | a^1)P(c^0 | b^0)P(d^1 | c^0) +$$

$$P(a^0)P(b^0 | a^0) + P(a^1)P(b^0 | a^1)P(c^1 | b^0)P(d^1 | c^1) +$$

$$P(a^0)P(b^1 | a^0) + P(a^1)P(b^1 | a^1)P(c^0 | b^1)P(d^1 | c^0) +$$

$$\left(P(a^0)P(b^1 | a^0) + P(a^1)P(b^1 | a^1)\right)P(c^1 | b^1)P(d^1 | c^1)$$

Now, replace these with symbols that we will compute only once and use them everywhere. Replacing $P(a^0)P(b^0 | a^0) + P(a^1) * P(b^0 | a^1)$ with $\tau_1(b^0)$ and $P(a^0)P(b^1 | a^0) + P(a^1) * P(b^1 | a^1)$ with $\tau_1(b^1)$, we get:

$$P(d^0) = \tau_1(b^0)P(c^0 | b^0)P(d^0 | c^0) +$$

$$\tau_1(b^0)P(c^1 | b^0)P(d^0 | c^1) +$$

$$\tau_1(b^1)P(c^0 | b^1)P(d^0 | c^0) +$$

$$\tau_1(b^1)P(c^1 | b^1)P(d^0 | c^1)$$

$$P(d^1) = \tau_1(b^0)P(c^0 | b^0)P(d^1 | c^0) +$$

$$\tau_1(b^0)P(c^1 | b^0)P(d^1 | c^1) +$$

$$\tau_1\left(b^1\right)P\left(c^0\mid b^1\right)P\left(d^1\mid c^0\right)+$$

$$\tau_1\left(b^1\right)P\left(c^1\mid b^1\right)P\left(d^1\mid c^1\right)+$$

Again, grouping common parts together, we get:

$$P\left(d^0\right)=\left(\tau_1\left(b^0\right)P\left(c^0\mid b^0\right)+\tau_1\left(b^1\right)P\left(c^0\mid b^1\right)\right)P\left(d^0\mid c^0\right)+$$

$$\left(\tau_1\left(b^0\right)P\left(c^1\mid b^0\right)+\tau_1\left(b^1\right)P\left(c^1\mid b^1\right)\right)P\left(d^0\mid c^1\right)$$

$$P\left(d^1\right)=\left(\tau_1\left(b^0\right)P\left(c^0\mid b^0\right)+\tau_1\left(b^1\right)P\left(c^0\mid b^1\right)\right)P\left(d^1\mid c^0\right)+$$

$$\left(\tau_1\left(b^0\right)P\left(c^1\mid b^0\right)+\tau_1\left(b^1\right)P\left(c^1\mid b^1\right)\right)P\left(d^1\mid c^1\right)$$

Now, replacing $\tau_1(b^0)*P(c^0\mid b^0)+\tau_1(b^1)*P(c^0\mid b^1)$ with $\tau_2(c^0)$ and replacing $\tau_1(b^0)*P(c^1\mid b^0)+\tau_1(b^1)*P(c^1\mid b^1)$ with $\tau_2(c^1)$, we get:

$$P\left(d^0\right)=\tau_2\left(c^0\right)P\left(d^0\mid c^0\right)+$$

$$\tau_2\left(c^1\right)P\left(d^0\mid c^1\right)$$

$$P\left(d^1\right)=\tau_2\left(c^0\right)P\left(d^1\mid c^0\right)+$$

$$\tau_2\left(c^1\right)P\left(d^1\mid c^1\right)$$

Notice how, instead of doing a summation over the complete product of $P(a)P(b\mid a)P(c\mid b)P(d\mid c)$ here, we did the summation over parts of it:

$$P(D)=\sum_a\sum_b\sum_c P(a)P(b\mid a)P(c\mid b)P(D\mid c)$$

$$P(D)=\sum_c P(D\mid c)\sum_b P(c\mid b)\sum_a P(a)P(b\mid a)$$

Here, we have been able to push the summations inside the equation because not all the terms in the equation have all the variables. So, only the terms $P(a)$ and $P(b\mid a)$ depend on A. So we can simply sum them over the values of A.

To make things clearer, let's see another run of variable elimination on the restaurant model:

Step	Variable eliminated	Factors involved	Intermediate factor	New factor
1	L	$\phi(L), \phi(L,C,Q), \phi(L,C,N)$	$\psi_1(L,C,Q,N)$	$\tau_1(C,Q,N)$
2	Q	$\phi(Q), \tau_1(C,Q,N)$	$\psi_2(C,N,Q)$	$\tau_2(C,N)$
3	C	$\tau_2(C,N)$	$\psi_3(C,N)$	$\tau_3(N)$
4	N	$\tau_3(N)$	$\psi_4(N)$	$\tau_4(\theta)$

This method helps us to significantly reduce the computation required to compute the probabilities. In this case, we just need to compute $\tau_1(B)$, which requires two multiplications and two additions, and $\tau_2(C)$, which requires four multiplications and two additions. We can then compute $P(D)$. Hence, we just need a total of 12 computations to compute $P(D)$. However, in the case of computing $P(D)$ from the joint probability distribution, we require 16 * 3 = 48 multiplications and 14 additions. Hence, we see that using variable elimination brings about huge improvement in the complexity of making the inference.

Now, let's see how to make the inference using variable elimination with pgmpy:

```
In [1]: from pgmpy.models import BayesianModel
In [2]: from pgmpy.inference import VariableElimination
In [3]: from pgmpy.factors import TabularCPD

# Now first create the model.
In [3]: restaurant = BayesianModel(
                            [('location', 'cost'),
                             ('quality', 'cost'),
                             ('cost', 'no_of_people'),
                             ('location', 'no_of_people')])
In [4]: cpd_location = TabularCPD('location', 2, [[0.6, 0.4]])
In [5]: cpd_quality = TabularCPD('quality', 3, [[0.3, 0.5, 0.2]])
In [6]: cpd_cost = TabularCPD('cost', 2,
                        [[0.8, 0.6, 0.1, 0.6, 0.6, 0.05],
                         [0.2, 0.1, 0.9, 0.4, 0.4, 0.95]],
                        ['location', 'quality'], [2, 3])
In [7]: cpd_no_of_people = TabularCPD(
                        'no_of_people', 2,
                        [[0.6, 0.8, 0.1, 0.6],
                         [0.4, 0.2, 0.9, 0.4]],
                        ['cost', 'location'], [2, 2])
```

```
In [8]: restaurant.add_cpds(cpd_location, cpd_quality,
                             cpd_cost, cpd_no_of_people)

# Creating the inference object of the model
In [9]: restaurant_inference = VariableElimination(restaurant)

# Doing simple queries over one or multiple variables.
In [10]: restaurant_inference.query(variables=['location'])
Out[10]: {'location': <Factor representing phi(location:2) at
                                               0x7fea25e02898>}
In [11]: restaurant_inference.query(
                        variables=['location', 'no_of_people'])
Out[11]: {'location': <Factor representing phi(location:2) at
                                               0x7fea25e02b00>,
          'no_of_people': <Factor representing phi(no_of_people:2)
                                            at 0x7fea25e026a0>}

# We can also specify the order in which the variables are to be
# eliminated. If not specified pgmpy automatically computes the
# best possible elimination order.
In [12]: restaurant_inference.query(variables=['no_of_people'],
              elimination_order=['location', 'cost',  'quality'])
Out[12]: {'no_of_people': <Factor representing phi(no_of_people:2)
                                          at 0x7fea25e02160>
```

We saw the case of making an inference when no condition was given. Now, let's take a case where some evidence is given; let's say we know that the cost of the restaurant is high and we want to compute the probability of the number of people in the restaurant. Basically, we want to compute $P(N|c^1)$. For this, we could simply use the probability theory and first compute $P(N,c^1)$ and then normalize this over N again to get $P(c^1)$ and then compute $P(N,c^1)$ as:

$$P(N|c^1) = \frac{P(N,c^1)}{P(c^1)}$$

Now, the question is, how do we compute $P(N|c^1)$? We first reduce all the factors involving C to c^1 and then do our normal variable elimination.

In pgmpy, we can simply pass another argument to the query method for evidence. Let's see how to find $P(N|c^1)$ using pgmpy:

```
# If we have some evidence for the network we can simply pass it
# as an argument to the query method in the form of
# {variable: state}
In [13]: restaurant_inference.query(variables=['no_of_people'],
                             evidence={'location': 1})
```

```
Out[13]: {'no_of_people': <Factor representing phi(no_of_people:2)
                                         at 0x7fea25e02588>}
In [14]: restaurant_inference.query(
                    variables=['no_of_people'],
                    evidence={'location': 1, 'quality': 1})
Out[14]: {'no_of_people': <Factor representing phi(no_of_people:2)
                                         at 0x7fea25e02d30>}
# In this case also we can simply pass the elimination order for
# the variables.
In [15]:restaurant_inference.query(
                    variables=['no_of_people'],
                    evidence={'location': 1},
                    elimination_order=['quality', 'cost'])
Out[15]: {'no_of_people': <Factor representing phi(no_of_people:2)
                                         at 0x7fea25e02eb8>}
```

Analysis of variable elimination

We have already seen that variable elimination is much more efficient for calculating probability distributions than normalizing and marginalizing the joint probability distribution. Now, let's do an exact analysis to find the complexity of variable elimination.

Let's start by putting the variable elimination algorithm in simple terms. In variable elimination, we start by choosing a variable X_i, then we compute the factor product ψ_j for all the factors involving that variable, and then eliminate that variable by summing it up, resulting in a new factor τ_i whose scope is $\{Scope(\psi_i) - X_i\}$.

Now, let's consider that we have a network with n variables and m factors. In the case of a Bayesian network, the number of CPDs will always be equal to the number of variables, therefore, $m = n$ for a Bayesian network. However, in the case of a Markov network, the number of factors can be more than the number of variables in the network. For simplicity, let's assume that we will be eliminating all the variables in the network.

In variable elimination, we have been performing just two types of operations, multiplication and addition. So, to find the overall complexity, let's start by counting these operations. For the multiplication step, we multiply each of the initial m factors and the intermediately formed n factors exactly once. So, the total number of multiplication steps would be $(m+n)N_i$, where N_i is the size of the intermediate factor ψ_j. Also, let's define $N_{max} = max_i N_i$. Therefore, $(m+n)N_i \leq (m+n)N_{max}$ will always be true. Now, if we calculate the total number of addition operations, we will be iterating over each of the ψ_j once, resulting in a total of nN_{max} addition operations. So, we see that the total number of operations for Variable Elimination comes out to be $(m+n)N_{max} + nN_{max}$. Hence, the complexity of the overall operation is $O(mN_{max})$, because $n \leq m$.

Now, let's try to analyze the variable elimination algorithm using the graph structure. We can treat a Bayesian model as a Markov model having undirected edges between all the variables in each of the CPD defining the parameters of the network. Now, let's try to see what happens when we run variable elimination over this network. We choose any variable X and multiply all other factors involving that factor X, resulting in a factor ψ with the scope $X \cup neigh(X)$. After this, we eliminate the variable X and have a resulting factor τ with scope $neigh(X) = Y$. Now, as we have a factor with scope Y, we need to have edges in the network between each of the variables in Y. So, we add extra edges to the network, which are known as fill edges. For the elimination of the next variable, we use this new network structure and perform similar operations on it.

Let's see an example on our late-for-school model showing the graph structure during the various steps of variable elimination:

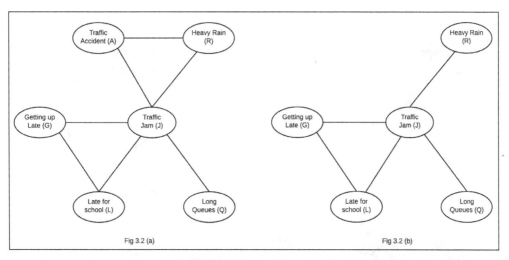

Fig 3.2 (a)

Fig 3.2 (b)

Fig 3.2(a): Initial state of the network

Fig 3.2(b): After eliminating Traffic Accident (A)

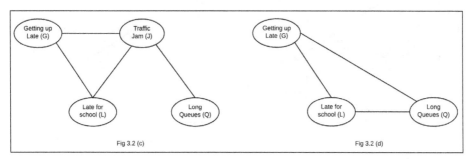

Fig 3.2 (c)

Fig 3.2 (d)

Fig 3.2(c): After eliminating Heavy Rain (R)

Fig 3.2(d): After eliminating Traffic Jam (J)

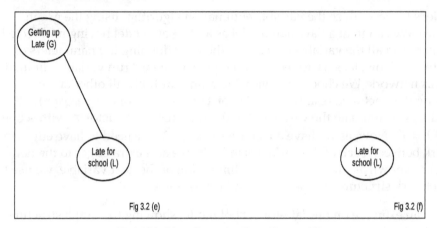

Fig 3.2 (e)   Fig 3.2 (f)

Fig 3.2(e): After eliminating Long Queues (Q)

Fig 3.2(f): After eliminating Getting Up Late (G)

An induced graph is also defined as the undirected graph constructed by the union of all the graphs formed in each step of variable elimination on the network. Fig 3.3 shows the induced graph for the preceding variable elimination:

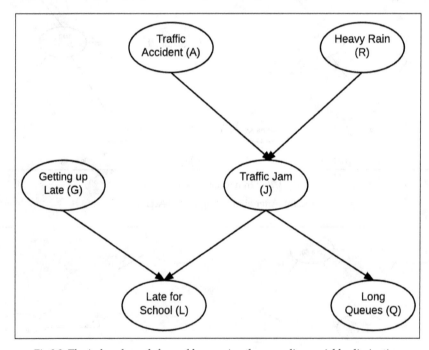

Fig 3.3: The induced graph formed by running the preceding variable elimination

We can also check the induced graph using `pgmpy`:

```
In [16]: induced_graph = restaurant_inference.induced_graph(
                ['cost', 'location', 'no_of_people', 'quality'])
In [17]: induced_graph.nodes()
Out[17]: ['location', 'quality', 'cost', 'no_of_people']
In [18]: induced_graph.edges()
Out[18]:
[('location', 'quality'),
 ('location', 'cost'),
 ('location', 'no_of_people'),
 ('quality', 'cost'),
 ('quality', 'no_of_people'),
 ('cost', 'no_of_people')]
```

Finding elimination ordering

In variable elimination, the order in which we eliminate the variables has a huge impact on the computational cost of running the algorithm. Let's look at the difference in the elimination ordering on the late-for-school model. The steps of variable elimination with the elimination order A, G, J, L, Q, R are shown in the following table:

Step	Variable eliminated	Factors involved	Intermediate factors	New factor
1	A	$\phi(A), \phi(J, A, R)$	$\psi_1(J, A, R)$	$\tau_1(J, R)$
2	G	$\phi(G), \phi(L, J, G)$	$\psi_2(J, L, G)$	$\tau_2(J, L)$
3	J	$\phi(Q, J), \tau_1(J, R), \tau_2(J, L)$	$\psi_3(Q, R, L, J)$	$\tau_3(Q, R, L)$
4	L	$\tau_3(Q, R, L)$	$\psi_4(Q, R, L)$	$\tau_4(Q, R)$
5	Q	$\tau_4(Q, R)$	$\psi_5(Q, R)$	$\tau_5(R)$
6	R	$\phi(R), \tau_5(R)$	$\psi_6(R)$	$\tau_6(\theta)$

The steps of variable elimination with the elimination order R, Q, L, J, G, A are shown in the following table:

Step	Variable eliminated	Factors involved	Intermediate factors	New factor
1	R	$\phi(R), \phi(J, A, R)$	$\psi_1(J, A, R)$	$\tau_i(J, A)$
2	Q	$\phi(Q, J)$	$\psi_2(Q, J)$	$\tau_2(J)$
3	L	$\phi(L, J, G)$	$\psi_3(L, J, G)$	$\tau_3(J, G)$
4	J	$\tau_1(J, A), \tau_2(J), \tau_3(J, G)$	$\psi_4(J, A, G)$	$\tau_4(A, G)$
5	G	$\phi(G), \tau_4(A, G)$	$\psi_5(A, G)$	$\tau_5(A)$
6	A	$\phi(A), \tau_5(A)$	$\psi_6(A)$	$\tau_6(\theta)$

For every intermediate factor, we add filled edges between all the variables in their scope, so we can say that every intermediate factor introduces a clique in the induced graph. Hence, the scope of every intermediate factor generated during the variable elimination process is a clique in the induced graph. Also, notice that every maximal clique in the induced graph is the scope of some intermediate factor generated during the variable elimination process. Therefore, having a larger maximal clique in the induced graph also means having a larger intermediate factor, which means higher computation cost.

Let's see a few definitions related to induced graphs:

- **Width**: This is defined as the number of nodes in the largest clique of the graph minus 1
- **Induced width**: This is defined as the width of an induced graph over some network, given an elimination ordering
- **Tree width**: The tree width of a network is defined as its minimal induced width

We have seen how the computation complexity of the variable elimination operation relates to the choice of elimination order, and how this relates to the tree width of the induced graph. A smaller tree width ensures a better complexity compared to an elimination order with a higher tree width. So, our problem has now been reduced to selecting an elimination order that keeps the tree width minimal.

Unfortunately, finding the elimination for the minimal tree width is NP -complete, so there is no easy way to find the complexity of the inference over a network by simply looking at the network structure. However, there are many other techniques that we can use to find good elimination orderings.

Using the chordal graph property of induced graphs

We define a graph as being chordal if it contains no cycle of length greater than three, and if there are no edges between two nonadjacent nodes of each cycle. In other words, every minimal cycle in a chordal graph is three in length.

Now, if you look carefully, you will see that every induced graph is a chordal graph. Also, the converse of this theorem holds, that every chordal graph on these variables corresponds to some elimination ordering. The proof of both of these theorems is beyond the scope of this book.

So, to find the elimination order, we use the maximum cardinality search algorithm. In this algorithm, we basically iterate $|\chi|$ times, and in each iteration, we try to find the variable with the largest number of marked variables and then mark that variable. This results in elimination ordering.

Minimum fill/size/weight/search

Another approach to find elimination ordering is to take the greedy approach and, in each step, select a variable that seems to be the best option for that step. So, for each iteration, we compute a cost function to eliminate each of the nodes and select the node that results in the minimum cost. Some of the cost criteria are as follows:

- **Min-neighbors**: The cost of a node is defined by the number of neighbors it has in the graph.
- **Min-weight**: The cost of a node is the product of the cardinality of its neighbors.
- **Min-fill**: The cost of node elimination is the number of edges that need to be added to the graph for the elimination of that node.
- **Weighted-min-fill**: The cost of node elimination is the sum of the weights of the edges that need to be added to the graph for its elimination. The weight of an edge is defined as the product of the weights of the nodes between which it lies.

Here, we have seen two different approaches to finding good elimination ordering. The second heuristic approach of going the greedy way doesn't seem to be a very good approach to get globally optimized elimination ordering. However, it turns out that it gives very good results in most of the cases as compared to our maximum cardinality search algorithm.

Belief propagation

In the previous section, we saw that the basic operation of the variable elimination algorithm is the manipulation of the factors. First, we create a factor ψ_j by multiplying existing factors. Then, we eliminate a variable in ψ_j to generate a new factor τ_i, which is then used to create another factor. From a different perspective, we can say that a factor ψ_j is a data structure, which takes messages τ_j generated by the other factor ψ_j, and generates a message ψ_j which is used by the other factor ψ_l.

Clique tree

Before we go into a detailed discussion of the belief propagation algorithm, let's discuss the graphical model that provides the basic framework for it, the clique tree, also known as the **junction tree**.

The clique tree (τ) is an undirected graph over a set of factors Φ, where each node represents a cluster of random variables and the edges connect the clusters, whose scope has a nonempty intersection. Thus, each edge between a pair of clusters C_i and C_j is associated with a sepset $S_{i,j} \subseteq C_i \cap C_j$. For each cluster C_i, we also define the cluster potential ψ_j, which is the factor representing all the variables present in it.

This can be generalized. Let's assume there is a variable X, such that $X \in C_i$ and $X \in C_j$. Then, X is also present in every cluster in the path between C_i and C_j in τ. This is known as the running intersection property. We can see an example in the following figure:

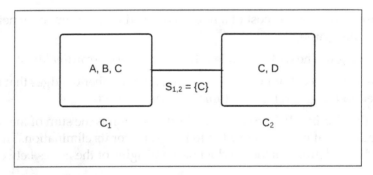

Fig 3.4: A simple cluster tree with clusters $C_1 = \{A, B, C\}$ and $C_2 = \{C, D\}$. The sepset associated with the edge is $S_{1,2} = \{C\}$.

In pgmpy, we can define a clique tree or junction tree in the following way:

```
# Firstly import JunctionTree class
In [1]: from pgmpy.models import JunctionTree
In [2]: junction_tree = JunctionTree()

# Each node in the junction tree is a cluster of random variables
# represented as a tuple
In [3]: junction_tree.add_nodes_from([('A', 'B', 'C'),
                                      ('C', 'D')])
In [4]: junction_tree.add_edge(('A', 'B', 'C'), ('C', 'D'))

In [5]: junction_tree.add_edge(('A', 'B', 'C'), ('D', 'E', 'F'))
        ValueError: No sepset found between these two edges.
```

 As discussed previously, the junction tree contains undirected edges only between those clusters whose scope has a non empty intersection. So, if we try to add any edge between two nodes whose scope has an empty intersection, it will raise ValueError.

Constructing a clique tree

In the previous section on variable elimination, we saw that an induced graph created by variable elimination is a chordal graph. The converse of it also holds true; that is, any chordal graph can be used as a basis for inference.

We previously discussed chordal graphs, triangulation techniques (the process of constructing a chordal graph that incorporates an existing graph), and their implementation in pgmpy. To construct a clique tree from the chordal graph, we need to find the maximal cliques in it. There are multiple ways of doing this. One of the simplest methods is the maximum cardinality search (which we discussed in the previous section) to obtain maximal cliques.

Then, these maximal cliques are assigned as nodes in the clique tree. Finally, to determine the edges of the clique tree, we use the maximum spanning tree algorithm. We build an undirected graph whose nodes are maximal cliques in H, where every pair of nodes C_i, C_j is connected by an edge whose weight is $|C_i \cap C_j|$. Then, by applying the maximum spanning tree algorithm, we find a tree in which the weight of edges is at maximum.

The cluster potential for each cluster in the clique tree can be computed as the product of the factors associated with each node of the cluster. For example, in the following figure, ψ_1 (the cluster potential associated with cluster (A, B, C) is computed as the product of $P(A)$, $P(C)$, and $P(B|A,C)$. To compute ψ_2 (the cluster potential associated with (B, D, E), we use $P(E|B,D)$, $P(B)$, and $P(D)$. $P(B)$ is computed by marginalizing $P(B|A,C)$ with respect to A and C.

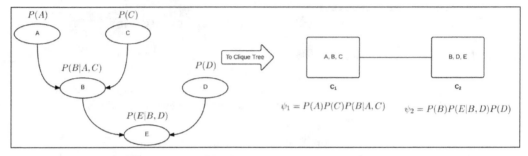

Fig 3.5: The cluster potential of the clusters present in the clique tree

These steps can be summarized as follows:

1. Triangulate the graph G over factor Φ to create a chordal graph H_Φ.

2. Find the maximal cliques in H_Φ and assign them as nodes to an undirected graph.

3. Assign weights to the edges between two nodes of the undirected graph as the numbers of elements in the sepset of the two nodes.

4. Construct the clique tree using the maximum spanning tree algorithm.

5. Compute the cluster potential for each cluster as the product of factors associated with the nodes present in the cluster.

In pgmpy, each graphical model class has a method called to_junction_tree(), which creates a clique tree (or junction tree) corresponding to the graphical model. Here's an example:

```
In [1]: from pgmpy.models import BayesianModel, MarkovModel
In [2]: from pgmpy.factors import TabularCPD, Factor

# Create a bayesian model as we did in the previous chapters
In [3]: model = BayesianModel(
                    [('rain', 'traffic_jam'),
                     ('accident', 'traffic_jam'),
                     ('traffic_jam', 'long_queues'),
                     ('traffic_jam', 'late_for_school'),
                     ('getting_up_late', 'late_for_school')])
```

```
In [4]: cpd_rain = TabularCPD('rain', 2, [[0.4], [0.6]])
In [5]: cpd_accident = TabularCPD('accident', 2, [[0.2], [0.8]])
In [6]: cpd_traffic_jam = TabularCPD(
                        'traffic_jam', 2,
                        [[0.9, 0.6, 0.7, 0.1],
                         [0.1, 0.4, 0.3, 0.9]],
                        evidence=['rain', 'accident'],
                        evidence_card=[2, 2])
In [7]: cpd_getting_up_late = TabularCPD('getting_up_late', 2,
                                [[0.6], [0.4]])
In [8]: cpd_late_for_school = TabularCPD(
                        'late_for_school', 2,
                        [[0.9, 0.45, 0.8, 0.1],
                         [0.1, 0.55, 0.2, 0.9]],
                        evidence=['getting_up_late',
                                'traffic_jam'],
                        evidence_card=[2, 2])
In [9]: cpd_long_queues = TabularCPD('long_queues', 2,
                                [[0.9, 0.2],
                                 [0.1, 0.8]],
                                evidence=['traffic_jam'],
                                evidence_card=[2])
In [10]: model.add_cpds(cpd_rain, cpd_accident,
                        cpd_traffic_jam, cpd_getting_up_late,
                        cpd_late_for_school, cpd_long_queues)
In [11]: junction_tree_bm = model.to_junction_tree()
In [12]: type(junction_tree_bm)
Out[12]: pgmpy.models.JunctionTree.JunctionTree

In [13]: junction_tree_bm.nodes()
Out[13]:
[('traffic_jam', 'getting_up_late', 'late_for_school'),
 ('traffic_jam', 'rain', 'accident'),
 ('traffic_jam', 'long_queues')]

In [14]: junction_tree_bm.edges()
Out[14]:
[(('traffic_jam', 'long_queues'),
  ('traffic_jam', 'late_for_school', 'getting_up_late')),
 (('traffic_jam', 'long_queues'), ('traffic_jam', 'rain',
'accident'))]
```

The `to_junction_tree()` method is available in `FactorGraph`, `MarkovModel` classes as well.

Message passing

Let's go back to the previous example of the Bayesian network for the late- for school- example:

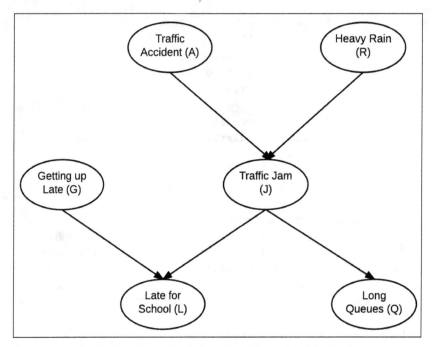

Fig 3.6: Bayesian network for a student being late for school.

In the previous section, we saw how to construct a clique tree for this Bayesian network. The following figure shows the clique tree for this network:

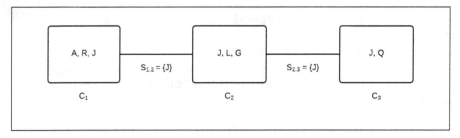

Fig 3.7: Clique tree constructed from the Bayesian network presented in Fig 3.3

As we discussed earlier, in the belief propagation algorithm, we would be considering factor ψ_j to be a computational data structure that would take a message $\tau_{j\to i}$ generated from a factor ψ_j, and produce $\tau_{i\to k}$, which can be further passed on to another factor ψ_k, and so on.

Let's go into the details of what each message term (τ_j and ψ) means. Let's begin with a very simple example of finding the probability of being late for school (L). To compute the probability of L, we need to eliminate all the random variables, such as accident (A), rain (R), traffic jam (J), getting up late (G), and long queues (Q). We can see that variables A and R are present only in cluster C_1 and Q is present only in C_3, but J is present in all three clusters, namely C_1, C_2, and C_3. So, both A and R can be eliminated from C_1 by just marginalizing ψ_1 with respect to A and R. Similarly, we could eliminate Q from ψ_3. However, to eliminate J, we can't just eliminate it from C_1, C_2, or ψ_3 alone. Instead, we need contributions from all three.

Eliminating the random variables A and R by marginalizing the cluster potential ψ_1 corresponding to C_1, we get the following:

$$\tau_{1\to2}(J) = \sum_A \sum_R \psi_1(A,R,J)$$

Similarly, marginalizing the cluster potential ψ_3 corresponding to C_3 with respect to Q, we get the following:

$$\tau_{3\to2}(J) = \sum_Q \psi_3(J,Q)$$

Now, to eliminate J and G, we must use $\tau_{1\to2}(J)$, $\tau_{3\to2}(J)$, and $\psi_2(J,L,G)$. Eliminating J and G, we get the following:

$$\phi(L) = \sum_G \sum_J \tau_{1\to2}(J)\tau_{3\to2}(J)\psi_2(J,L,G)$$

From the perspective of message passing, we can see that ψ_1 produces an output message $\tau_{1\to2}$. Similarly, ψ_3 produces a message $\tau_{1\to2}$. These messages are then used as input messages for ψ_2 to compute the belief for a cluster C_2. Belief for a cluster C_i is defined as the product of the cluster potential ψ_j with all the incoming messages to that cluster. Thus:

$$\beta_2(J,L,G) = \tau_{1\to2}(J)\tau_{3\to2}(J)\psi_2(J,L,G)$$

So, we can re-frame the following equation:

$$\phi(L) = \sum_G \sum_J \tau_{1\to2}(J)\tau_{3\to2}(J)\psi_2(J,L,G)$$

It can be re-framed as follows:

$$\phi(L) = \sum_G \sum_J \beta_2(J,L,G)$$

Fig 3.8 shows message propagation from clusters C_1 and C_3 to cluster C_2:

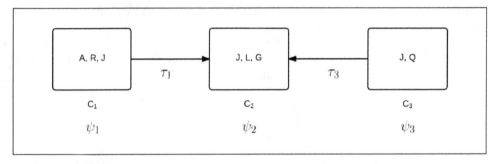

Fig 3.8: Message propagation from clusters C_1 and C_3 to cluster C_2:

Now, let's consider another example, where we compute the probability of long queues (Q). We have to eliminate all the other random variables, except Q. Using the same approach as discussed earlier, first marginalize ψ_1 with respect to A and R, and compute τ_1 as follows:

$$\tau_{1\to2}(J) = \sum_A \sum_R \psi_1(A,R,J)$$

As discussed earlier, to eliminate the variable J, we need contributions from C_1, C_2, and C_3, so we can't simply eliminate J from ψ_2. The other two random variables L and G are only present in C_2, so we can easily eliminate them from C_2. However, to eliminate L and G from C_2, we can't simply marginalize ψ_2. We have to consider the contribution of $\tau_{1\to2}$ (the outgoing message from C_1) as well, because J was present in both the clusters C_1 and C_2. Thus, eliminating L and G would create τ_2 as follows:

$$\tau_{2\to3} = \sum_L \sum_G \tau_{1\to2}(J)\psi_2(J,L,G)$$

Finally, we can eliminate J by marginalizing the following belief β_3 of C_3:

$$\beta_3(J,Q) = \psi_3(J,Q)\tau_{2\rightarrow3}(J)$$

We can eliminate it as follows:

$$\phi(Q) = \sum_J \beta_3(J,Q)$$

Fig 3.9 shows a message passing from C_1 to C_2 and C_2 to C_3:

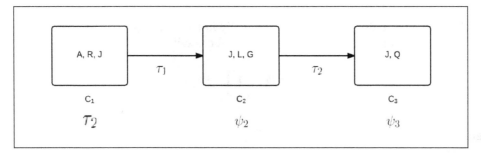

Fig 3.9: Message passing from C_1 to C_2 and C_2 to C_3

In the previous examples, we saw how to perform variable elimination in a clique tree. This algorithm can be stated in a more generalized form. We saw that variable elimination in a clique tree induces a directed flow of messages between the clusters present in it, with all the messages directed towards a single cluster, where the final computation is to be done. This final cluster can be considered as the root. In our first example, the root was C_2, while in the second example, it was C_3. The notions of directionality and root also create the notions of upstream and downstream. If C_i is on the path from C_j to the root, then we can say that C_i is upstream from C_j, and C_j is downstream from C_i:

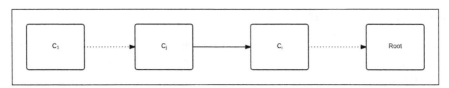

Fig 3.10: C_i is upstream from C_j and C_j is downstream from C_i

We also saw in the second example that C_2 was not able to send messages to C_3 until it received the message from C_1, as the generation of $\tau_{2\to3}$ also depends on $\tau_{1\to2}$. This introduces the notion of a ready cluster. A cluster is said to be ready to transmit messages to its upstream cliques if it has received all the incoming messages from its downstream cliques.

The message C_3 from the cluster j to the cluster i can be defined as the factor formed by the following sum product message passing computation:

$$\tau_{j\to i} = \sum_{C_j - S_{i,j}} \psi_j \prod_{k \in Neighbor(j) - \{i\}} \tau_{k \to j}$$

We can now define the terms belief β_1 of a cluster C_i. It is defined as the product of all the incoming messages $\tau_{k\to i}$ from its neighbors with its own cluster potential:

$$\beta_i = \psi_i \prod_{k \in Neighbor(i)} \tau_{k \to i}$$

Here, j is the upstream from i.

All these discussions for running the algorithm can be summarized in the following steps:

1. Identify the root (this is the cluster where the final computation is to be made).
2. Start with the leaf nodes of the tree. The output message of these nodes can be computed by marginalizing its belief. The belief for the leaf node would be its cluster potential as there would be no incoming message.
3. As and when the other clusters of the clique tree become ready, compute the outgoing message and propagate them upstream.
4. Repeat step 3 until the root node has received all the incoming messages.

Clique tree calibration

In the previous section, we discussed how to compute the probability of any variable using belief propagation. Now, let's look at the broader picture. What if we wanted to compute the probability of more than one random variable? For example, say we want to know the probability of long queues as well as a traffic jam. One naive solution would be to do a belief propagation in the clique tree by considering each cluster as a root. However, can we do better?

Consider the previous two examples we have discussed. The first one had C_2 as the root, while the other had C_3. We saw that in both cases, message computed from the cluster C_1 to the cluster C_2 (that is $\tau_{1\to2}$) is the same, irrespective of the root node. Generalizing this, we can conclude that the message $\tau_{j\to i}$ from the cluster C_j to the cluster C_i will be the same as long as the root is on the C_i side and vice versa. Thus, for a given edge in the clique tree between two clusters C_i and $C_{j'}$, we have only two messages to compute, depending on the directionality of the edges ($\tau_{i\to j}$ and $\tau_{j\to i}$). For a given clique tree with c clusters, we have $c-1$ edges between these clusters. Thus, we only need to compute $2(c-1)$ messages.

As we have seen in the previous section, a cluster can propagate a message upstream as soon as it is ready, that is, when it has received all the incoming messages from downstream. So, we can compute both messages for each edge asynchronously. This can be done in two phases, one being an upward pass and the other being a downward pass. In the upward pass (Fig 3.11), we consider a cluster as a root and send all the messages to the root. Once the root has all the messages, we can compute its belief. For the downward pass (Fig 3.12), we can compute appropriate messages from the root to its children using its belief. This phase will continue until there is no message to be transmitted, that is, until we have reached the leaf nodes. This is shown in Fig 3.11:

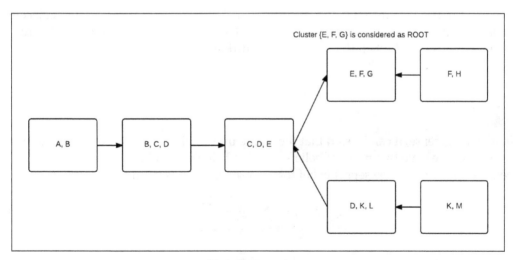

Fig 3.11: Upward pass

Fig 3.11 shows an upward pass where cluster {E, F, G} is considered as the root node. All the messages from the other nodes are transmitted towards it.

The following figure shows a downward pass where the appropriate message from the root is transmitted to all the children. This will continue until all the leaves are reached:

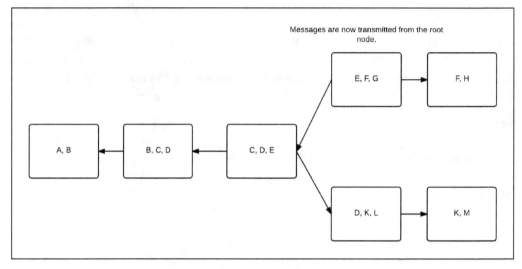

Fig 3.12: Downward pass

When both, the upward pass and the downward pass are completed, all the adjacent clusters in the clique tree are said to be calibrated. Two adjacent clusters i and j are said to be calibrated when the following condition is satisfied:

$$\sum_{C_i - S_{i,j}} \beta_i = \sum_{C_j - S_{i,j}} \beta_j$$

In a broader sense, it can be said that the clique tree is calibrated. When a clique tree is calibrated, we have two types of beliefs, the first being cluster beliefs and the second being sepset beliefs. The sepset belief for a sepset $S_{i,j}$ can be defined as follows:

$$\mu_{i,j}\left(S_{i,j}\right) = \sum_{C_i - S_{i,j}} \beta_i = \sum_{C_j - S_{i,j}} \beta_j$$

Message passing with division

Until now, we have viewed message passing in the clique tree from the perspective of variable elimination. In this section, we will see the implementation of message passing from a different perspective, that is, from the perspective of clique beliefs and sepset beliefs. Before we go into details of the algorithm, let's discuss another important operation on the factor called **factor division**.

Factor division

A factor division between two factors $\phi_1(X, Y)$ and $\phi_2(Y)$, where both X and Y are disjoint sets, is defined as follows:

$$\psi(X, Y) = \frac{\phi_1(X, Y)}{\phi_2(Y)}$$

Here, we define $\frac{0}{0} = 0$. This operation is similar to the factor product, except that we divide instead of multiplying. Further, unlike the factor product, we can't divide factors not having any common variables in their scope. For example, consider the following two factors:

a	b	$\Phi_1(a,b)$
a_0	b_0	0
a_0	b_1	1
a_0	b_2	2
a_1	b_0	3
a_1	b_1	4
a_1	b_2	5

b	$\Phi_2(b)$
b_0	0
b_1	1
b_2	2

Dividing $\Phi_1(a,b)$ by $\Phi_2(b)$, we get the following:

a	b	$\Psi(a,b)$
a_0	b_0	0
a_0	b_1	1
a_0	b_2	1
a_1	b_0	0
a_1	b_1	4
a_1	b_2	2.5

In pgmpy, factor division can implemented as follows:

```
In [1]: from pgmpy.factors import Factor
In [2]: phi1 = Factor(['a', 'b'], [2, 3], range(6))
In [3]: phi2 = Factor(['b'], [3], range(3))
In [4]: psi = phi1 / phi2
In [5]: print(psi)
```

a	b	phi(a,b)
a_0	b_0	0.0000
a_0	b_1	1.0000
a_0	b_2	1.0000
a_1	b_0	0.0000
a_1	b_1	4.0000
a_1	b_2	2.5000

Let's go back to our original discussion regarding message passing using division. As we saw earlier, for any edge between clusters C_i and C_j, we need to compute two messages $\tau_{i \rightarrow j}$ and $\tau_{j \rightarrow i}$. Let's assume that the first message was passed from C_i to C_j, that is, C_j. So, a return message from C_j to C_i would only be passed when C_j has received all the messages from its neighbors.

Once C_j has received all the messages from its neighbors, we can compute its belief β_j as follows:

$$\beta_j = \psi_j \prod_{k \in Neighbor(j)} \tau_{k \rightarrow j}$$

In the previous section, we also saw that the message from C_j to C_i can be computed as follows:

$$\tau_{j \rightarrow i} = \sum_{C_j - S_{i,j}} \psi_j \prod_{k \in Neighbor(j) - \{i\}} \tau_{k \rightarrow j}$$

From the preceding mathematical formulation, we can deduce that the belief of C_j, that is, β_j, can't be used to compute the message from C_j to C_i as it would already include the message from C_i to C_j in it:

$$\beta_j = \tau_{i \rightarrow j} \psi_j \prod_{k \in Neighbor(j) - \{i\}} \tau_{k \rightarrow j}$$

That is:

$$\beta_j = \tau_{i \rightarrow j} \tau_{j \rightarrow i}$$

Thus, from the preceding equation, we can conclude that the message from C_j to C_i can be computed by simply dividing the final belief of C_j, that is, β_j, with the message from C_i to C_j, that is, $\tau_{i \rightarrow j}$:

$$\tau_{j \rightarrow i} = \frac{\beta_j}{\tau_{i \rightarrow j}}$$

Finally, the message passing algorithm using this process can be summarized as follows:

1. For each cluster C_i, initialize the initial cluster belief β_1 as its cluster potential ψ_j and sepset potential between adjacent clusters C_i and C_j, that is, $\mu_{i,j}$ as 1.

2. In each iteration, the cluster belief β_1 is updated by multiplying it with the message from its neighbors, and the sepset potential $i - j$ is used to store the previous message passed along the edge $(i - j)$, irrespective of the direction of the message.

3. Whenever a new message is passed along an edge, it is divided by the old message to ensure that we don't count this message twice (as we discussed earlier).

   Steps 2 and 3 can formally be stated in the following way for each iteration:

$$\sigma_{i \to j} = \sum_{C_i - S_{i,j}} \beta_i$$

4. Here, we marginalize the belief to get the message passed. However, as we discussed earlier, this message will include a message from C_j to C_i in it, so divide it by the previous message stored in $i - j$:

$$\tau_{i \to j} = \frac{\sigma_{i \to j}}{\mu_{i,j}}$$

5. Update the belief by multiplying it with the message from its neighbors:

$$\beta_j \leftarrow \beta_j . \tau_{i \to j}$$

6. Update the sepset belief:

$$\mu_{i,j} \leftarrow \sigma_{i \to j}$$

7. Repeat steps 2 and 3 until the tree is calibrated for each adjacent edge $(i - j)$:

$$\mu_{i,j}\left(S_{i,j}\right) = \sum_{C_i - S_{i,j}} \beta_i = \sum_{C_j - S_{i,j}} \beta_j$$

As this algorithm updates the belief of a cluster using the beliefs of its neighbors, we call it the belief update message passing algorithm. It is also known as the Lauritzen-Spiegelhalter algorithm.

In pgmpy, this can be implemented as follows:

```
In [1]: from pgmpy.models import BayesianModel
In [2]: from pgmpy.factors import TabularCPD, Factor
In [3]: from pgmpy.inference import BeliefPropagation

# Create a bayesian model as we did in the previous chapters
In [4]: model = BayesianModel(
                    [('rain', 'traffic_jam'),
                     ('accident', 'traffic_jam'),
                     ('traffic_jam', 'long_queues'),
                     ('traffic_jam', 'late_for_school'),
                     ('getting_up_late', 'late_for_school')])

In [5]: cpd_rain = TabularCPD('rain', 2, [[0.4], [0.6]])
In [6]: cpd_accident = TabularCPD('accident', 2, [[0.2], [0.8]])
In [7]: cpd_traffic_jam = TabularCPD('traffic_jam', 2,
                            [[0.9, 0.6, 0.7, 0.1],
                             [0.1, 0.4, 0.3, 0.9]],
                            evidence=['rain',
                                      'accident'],
                            evidence_card=[2, 2])
In [8]: cpd_getting_up_late = TabularCPD('getting_up_late', 2,
                            [[0.6], [0.4]])
In [9]: cpd_late_for_school = TabularCPD(
                    'late_for_school', 2,
                    [[0.9, 0.45, 0.8, 0.1],
                     [0.1, 0.55, 0.2, 0.9]],
                    evidence=['getting_up_late','traffic_jam'],
                    evidence_card=[2, 2])
In [10]: cpd_long_queues = TabularCPD('long_queues', 2,
                            [[0.9, 0.2],
                             [0.1, 0.8]],
                            evidence=['traffic_jam'],
                            evidence_card=[2])

In [11]: model.add_cpds(cpd_rain, cpd_accident,
                    cpd_traffic_jam, cpd_getting_up_late,
                    cpd_late_for_school, cpd_long_queues)
```

```
In [12]: belief_propagation = BeliefPropagation(model)

# To calibrate the clique tree, use calibrate() method
In [13]: belief_propagation.calibrate()

# To get cluster (or clique) beliefs use the corresponding getters
In [14]: belief_propagation.get_clique_beliefs()
Out[14]:
{('traffic_jam', 'late_for_school', 'getting_up_late'): <Factor
representing phi(getting_up_late:2, late_for_school:2, traffic_jam:2)
at 0x7f565ee0db38>,
 ('traffic_jam', 'long_queues'): <Factor representing phi(long_
queues:2, traffic_jam:2) at 0x7f565ee0dc88>,
 ('traffic_jam', 'rain', 'accident'): <Factor representing phi(rain:2,
accident:2, traffic_jam:2) at 0x7f565ee0d4a8>}

# To get the sepset beliefs use the corresponding getters
In [15]: belief_propagation.get_sepset_beliefs()
Out[15]: {frozenset({('traffic_jam', 'long_queues'),
                     ('traffic_jam', 'rain', 'accident')}): <Factor
representing phi(traffic_jam:2) at 0x7f565ee0def0>,
         frozenset({('traffic_jam', 'late_for_school',
'getting_up_late'),
                   ('traffic_jam', 'long_queues')}): <Factor
representing phi(traffic_jam:2) at 0x7f565ee0dc18>}
```

Querying variables that are not in the same cluster

In the previous section, we saw how to compute the probability of variables present in the same cluster. Now, let's consider a situation where we want to compute the probability of both being late for school (L) and long queues (Q). These two variables are not present in the same cluster. So, to compute their probabilities, one naive approach would be to force our clique tree to have these two variables in the same cluster. However, this clique tree is not the optimal one, hence it would negate all the advantages of the belief propagation algorithm. The other approach is to perform variable elimination over the calibrated clique tree.

The algorithm for performing queries of variables not present in same cluster can be summarized as follows:

1. Select a subtree $\tau *$ of the calibrated clique tree τ , such that the query variable $Y \subseteq scope[\tau *]$. Let Φ be a set of factors on which variable elimination is to be performed. Select a cluster of the clique tree $\tau *$ as the root node and add its belief to Φ for each node in the clique tree Φ except the root node.

$$\phi = \frac{\beta_i}{\mu_i Parent(i)}$$

2. Add it to Φ. Let Z be a random set of random variables present in Φ, except for the query variables. Perform variable elimination on the set of factors Φ with respect to the variables Z.

In pgmpy, this can be implemented as follows:

```
In [15]: belief_propagation.query(
                        variables=['no_of_people'],
                        evidence={'location': 1, 'quality': 1})
Out[15]: {'no_of_people': <Factor representing phi(no_of_people:2)
                                     at 0x7f565ee0def0>
```

MAP inference

Until now, we have been doing conditional probability queries only on the model. However, sometimes, rather than knowing the probability of some given states of variables, we might be interested in finding the states of the variables corresponding to the maximum probability in the joint distribution. This type of problem often arises when we want to predict the states of variables in our model, which is our general machine learning problem. So, let's take the example of our restaurant model. Let's assume that for some restaurant we know of, the quality is good, the cost is low, and the location is good, and we want to predict the number of people coming to the restaurant. In this case, rather than querying for the probabilities of states of the number of people, we would like to query for the state that has the highest probability, given that the quality is good, the cost is low, and the location is good. Similarly, in the case of speech recognition, given a signal, we are interested in finding the most likely utterance rather than the probability of individual phonemes.

Putting the MAP problem more formally, we are given a distribution $P_\phi(\chi)$ defined by a set Φ and an unnormalized $\tilde{P}_\phi(\chi)$, and we want to find an assignment ξ whose probability is at maximum:

$$\xi^{map} = argmax_\xi P_\Phi(\chi) \quad \xi^{map} = argmax_\xi \frac{1}{Z}\tilde{P}_\Phi(\chi)$$

In the earlier equation, we used the unnormalized distribution to compute ξ^{map} as it helps us avoid computing the full distribution, because computing the partition function Z requires all the values of the distribution. Overall, the MAP problem is to find the assignment ξ for which $\tilde{P}_\phi(\chi)$ is at maximum.

A number of algorithms have been proposed to find the most likely assignment. Most of these use local maximum assignments and graph structures to find the global maximum likely assignment.

We define the max-marginal of a function f relative to a set of variables Y as follows:

$$MaxMarg_f(y) = \max_{\xi(Y)=y} f(\xi)$$

In simple words, $MaxMarg_{\tilde{P}_\Phi}(Y)$ returns the unnormalized probability value of the most likely assignment in $\tilde{P}_\Phi(Y)$. Most of the algorithms work on first computing this set of local max-marginals, that is $\{MaxMarg_f(X_i)\} x_i \in \chi$, and then use this to compute the global maximum assignment, as we will see in the next sections.

MAP using variable elimination

Let's start with a very basic example of a network $A \rightarrow B$, as shown in the following figure:

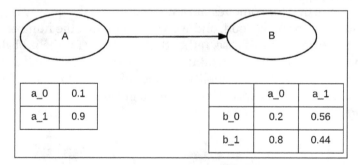

a_0	0.1
a_1	0.9

	a_0	a_1
b_0	0.2	0.56
b_1	0.8	0.44

Fig 3.13: Basic Bayesian network with two variables

For MAP, we want to compute the following:

$$\max_{a,b} P(a,b) = \max_{a,b} P(a)P(b\,|\,a) = \max_{a} \max_{b} P(a)P(b\,|\,a)$$

If we consider any particular assignment a for the variable A, we have the following:

$$\max_{a,b} P(a,b) = \max_{b} P(a)P(b\,|\,a)$$

So, for any given assignment of A, we have to select the assignment of B for which $P(b\,|\,a)$ is at maximum. We also have to select the maximum assignment of B as any given assignment of A doesn't guarantee that it would be the global maximum. Therefore, we need to check the values for each assignment of A.

Now, let's try to find the MAP assignment for the network in the Fig 3.13. Assuming the assignment from A to a_0, let's define $\phi(a^0) = \max_{b} P(b\,|\,a^0) = 0.8$ and similarly, $\phi(a^1) = \max_{b} P(b\,|\,a^1) = 0.56$. Now, let's compute the max-marginal over A:

$$\max_{a} P(a)\phi(a) = \max[0.1*0.8, 0.9*0.44] = 0.396$$

Factor maximization

For MAP queries in graphical models, we introduce another operation on factors called **maximization**.

Let X be a set of variables, $Y \notin X$ a variable, and $\phi(X,Y)$ a factor. We define factor maximization of Y in $\phi(X,Y)$ to be a factor ψ over the variables X such that the following occurs:

$$\psi(X) = \max_{Y} \phi(X,Y)$$

Let's take an example of factor maximization to make this clearer:

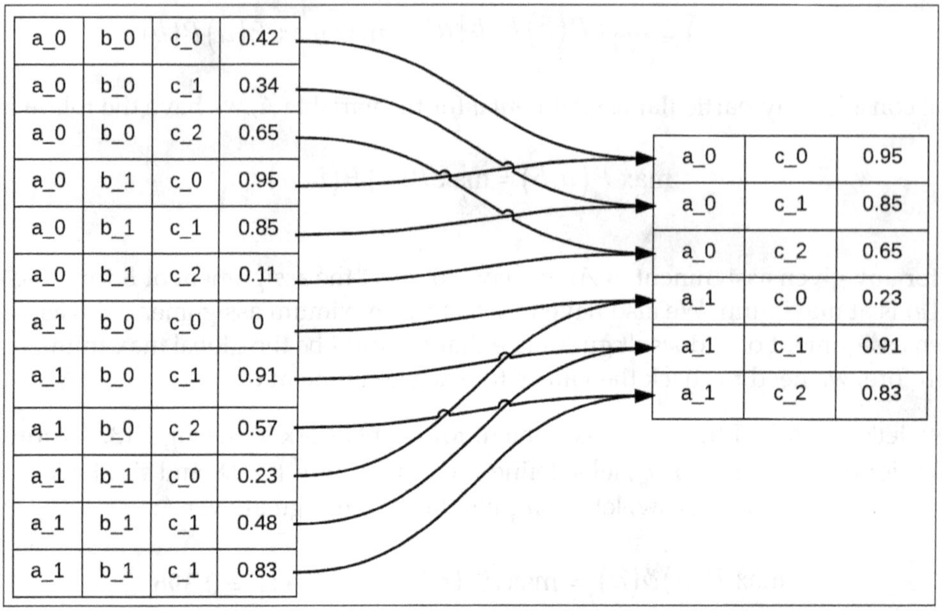

a_0	b_0	c_0	0.42
a_0	b_0	c_1	0.34
a_0	b_0	c_2	0.65
a_0	b_1	c_0	0.95
a_0	b_1	c_1	0.85
a_0	b_1	c_2	0.11
a_1	b_0	c_0	0
a_1	b_0	c_1	0.91
a_1	b_0	c_2	0.57
a_1	b_1	c_0	0.23
a_1	b_1	c_1	0.48
a_1	b_1	c_1	0.83

a_0	c_0	0.95
a_0	c_1	0.85
a_0	c_2	0.65
a_1	c_0	0.23
a_1	c_1	0.91
a_1	c_2	0.83

Fig 3.14: Factor maximization of variable B from a factor $\phi(A,B,C)$

Therefore, in the preceding example of the A -> B network, we had $\phi(A)=\max_{B}P(B|A)$. Also, another important property of maximization is that it can be inserted in equations if some of the factors don't involve the variable over which the maximization is being performed. More formally, for a variable $Y \notin Scope\ \phi_1$:

$$\max_{B}\left(\phi_1 * \phi_2\right)=\phi_1 * \max_{B}\phi_2$$

This is a very important property of maximization as it allows us to push the maximization operation inside equations, as we used to push summation in the case of the variable elimination operation. This avoids the full joint distribution and allows us to operate on much smaller factors.

Let's now try a sample run of the algorithm on the late-for-school model:

Step	Variable eliminated	Factors used	Intermediate factor	New factor
1	A	$\phi_A(A), \phi_J(J,A,R)$	$\psi_1(J,A,R)$	$\tau_1(J,R)$
2	J	$\phi_Q(Q,J), \phi_L(L,J,G), \tau_1(J,R)$	$\psi_2(Q,L,R,G,J)$	$\tau_2(Q,L,R,G)$
3	R	$\phi_R(R), \tau_2(Q,L,R,G)$	$\psi_3(Q,L,R,G)$	$\tau_3(Q,L,G)$
4	Q	$\tau_3(Q,L,G)$	$\psi_4(Q,L,G)$	$\tau_4(L,G)$
5	G	$\phi_G(G), \tau_4(L,G)$	$\psi_5(L,G)$	$\tau_5(L)$
6	L	$\tau_5(L)$	$\psi_6(L)$	$\tau_6(\theta)$

We can clearly see that the max-marginal operation is very similar to the variable elimination we performed. The only difference is that rather than marginalizing the intermediate factor over the variable to be eliminated, we maximize over the variable to be eliminated.

We can compute the max-marginal over networks using `pgmpy`:

```
In [1]: from pgmpy.models import BayesianModel
In [2]: from pgmpy.factors import TabularCPD
In [3]: from pgmpy.inference import VariableElimination

# Constructing the model
In [4]: model = BayesianModel(
                    [('rain', 'traffic_jam'),
                     ('accident', 'traffic_jam'),
                     ('traffic_jam', 'long_queues'),
                     ('traffic_jam', 'late_for_school'),
                     ('getting_up_late', 'late_for_school')])
In [5]: cpd_rain = TabularCPD('rain', 2, [[0.4], [0.6]])
In [6]: cpd_accident = TabularCPD('accident', 2, [[0.2], [0.8]])
In [7]: cpd_traffic_jam = TabularCPD(
                    'traffic_jam', 2,
                    [[0.9, 0.6, 0.7, 0.1],
                        [0.1, 0.4, 0.3, 0.9]],
                    evidence=['rain',
                              'accident'],
```

```
                                            evidence_card=[2, 2])
In [8]: cpd_getting_up_late = TabularCPD('getting_up_late', 2,
                                           [[0.6], [0.4]])
In [9]: cpd_late_for_school = TabularCPD(
                    'late_for_school', 2,
                        [[0.9, 0.45, 0.8, 0.1],
                         [0.1, 0.55, 0.2, 0.9]],
                        evidence=['getting_up_late', 'traffic_jam'],
                        evidence_card=[2, 2])
In [10]: cpd_long_queues = TabularCPD('long_queues', 2,
                                        [[0.9, 0.2],
                                         [0.1, 0.8]],
                                        evidence=['traffic_jam'],
                                        evidence_card=[2])
In [11]: model.add_cpds(cpd_rain, cpd_accident,
                        cpd_traffic_jam, cpd_getting_up_late,
                        cpd_late_for_school, cpd_long_queues)

# Calculating max marginals
In [12]: model_inference = VariableElimination(model)
In [13]: model_inference.max_marginal(
                            variables=['late_for_school'])
Out[13]: 0.5714285714285714
In [14]: model_inference.max_marginal(
                variables=['late_for_school', 'traffic_jam'])
Out[14]: 0.40547815820543098

# For any evidence in the network we can simply pass the evidence
# argument which is a dict of the form of {variable: state}
In [15]: model_inference.max_marginal(
                            variables=['late_for_school'],
                            evidence={'traffic_jam': 1})
Out[15]: 0.5714285714285714
In [16]: model_inference.max_marginal(
                            variables=['late_for_school'],
                            evidence={'traffic_jam': 1,
                                      'getting_up_late': 0})
Out[16]: 0.8000000000000004
In [17]: model_inference.max_marginal(
                    variables=['late_for_school','long_queues'],
                    evidence={'traffic_jam': 1,
                              'getting_up_late': 0}
Out[17]: 0.6399999999999999
```

```
# Again as in the case of VariableEliminaion we can also pass the
# elimination order of variables for MAP queries. If not specified
# pgmpy automatically computes the best elimination order for the
# query.
In [18]: model_inference.m_marginal(
                variables=['late_for_school'],
                elimination_order=['traffic_jam',
                                'getting_up_late', 'rain',
                                'long_queues', 'accident'])
Out[18]: 0.5714285714285714
In [19]: model_inference.max_marginal(
                variables=['late_for_school'],
                evidence={'accident': 1},
                elimination_order=['traffic_jam',

                                'getting_up_late',
                                'rain', 'long_queues'])
Out[19]: 0.57142857142857129
```

MAP using belief propagation

In the previous section, we discussed the MAP variable elimination algorithm. In the same way that we extended the sum-product variable elimination algorithm for the clique tree and ended up on the belief propagation algorithm, we can perform MAP using the belief propagation. In cases where variable elimination can be computationally intractable, belief propagation has a clear advantage.

The procedure for belief propagation remains the same as discussed in the case of the sum-product. The only thing that changes is the message that is passed between the two clusters C_i and C_j. Earlier, we used to compute messages from C_j to C_i, that is $\tau_{j \to i}$, as follows:

$$\tau_{j \to i} = \sum_{C_j - S_{i,j}} \psi_j \prod_{k \in Neighbor(j) - \{i\}} \tau_{k \to j}$$

However, now, instead of summing out the variables $C_j - S_{i,j}$, we will maximize with respect to them. Thus, the message in the case of MAP belief propagation can be formulated as follows:

$$\tau_{j \to i} = \max_{Cj - S_{i,j}} \psi_j \prod_{k \in Neighbour(j) - \{i\}} \tau_{k \to j}$$

When both, the upward pass and the downward pass of the messages are complete, all the adjacent clusters of the tree are said to be max-calibrated. At max-calibration, for any two adjacent clusters C_i and C_j, we have the following:

$$\mu_{i,j}\left(S_{i,j}\right) = \max_{C_i - S_{i,j}} \beta_i = \max_{C_j - S_{i,j}} \beta_j$$

A clique tree is said to be max-calibrated when all the adjacent edges are max-calibrated.

Finding the most probable assignment

In the previous section, we computed the maximum unnormalized probability value, but for MAP, we need to compute the states of the variables corresponding to the one in which this value occurs. Taking our earlier example of the network $A \rightarrow B$, we first computed $\max_{b} P(b \mid a)$, but the state of the variable B for which $P(b \mid a)$ gives the maximum value also depends on the state of the variable A. So, we will first need to compute $\max_{b} P(b \mid a)$ and then compute the state of B accordingly. So, from the CPDs of the network, we can see that $\max P(a)$. Now we will look for $\max_{a} P(a) = a^1$, which gives us $\max_{b} P(b \mid a^1)$. Hence, we get the maximum values corresponding to b_0 and a_0.

Also, the computational cost of this operation is not high, as we are simply doing another pass over the factors that have already been computed. Hence, the cost would be linear in the number of variables in the network.

Now, let's continue the previous code example and do some map queries over the networks using pgmpy:

```
In [20]: model_inference.map_query(variables=['late_for_school'])
Out[20]: {'late_for_school': 0}
In [21]: model_inference.map_query(variables=['late_for_school',
                                              'accident'])
Out[21]: {'accident': 1, 'late_for_school': 0}


# Again we can pass the evidence to the query using the evidence
# argument in the form of {variable: state}.
In [22]: model_inference.map_query(variables=['late_for_school'],
                                   evidence={'accident': 1})
Out[22]: {'late_for_school': 0}
In [23]: model_inference.map_query(variables=['late_for_school'],
                                   evidence={'accident': 1,
                                             'rain': 1})
Out[23]: {'late_for_school': 0}
```

```
# Also in the case of MAP queries we can specify the elimination
# order of the variables. But if the elimination order is not
# specified pgmpy automatically computes the best elimination
# order for the query.
In [24]: model_inference.map_query(
                      variables=['late_for_school'],
                      elimination_order=['accident', 'rain',
                                         'traffic_jam',
                                         'getting_up_late',
                                         'long_queues'])
Out[24]: {'late_for_school': 0}
In [25]: model_inference.map_query(
                      variables=['late_for_school'],
                  evidence={'accident': 1},
                  elimination_order=['rain',
                                     'traffic_jam',
                                     'getting_up_late',
                                     'long_queues'])
Out[25]: {'late_for_school': 0}

# Similarly MAP queries can be done for belief propagation as well.
In [26]: belief_propagation.map_query(['late_for_school'],
                                      evidence={'accident': 1})
Out[26]: {'late_for_school': 0}
```

Predictions from the model using pgmpy

In the previous sections, we have seen various algorithms to computing conditional distributions and learnt how to do MAP queries on the models. A MAP query is essentially a way to predict the states of variables, given the states of other variables. In a real-life problem, we are given some data with which we try to create a model for our problem. Then, using this trained model, we try to predict the states of variables for some new data point. This is the process with which we approach our supervised learning problems in machine learning.

Now, to design the models, we need to create CPDs or factors, add them to the base model, create an inference object, and then do MAP queries over it for new data points to predict variable states. This whole process is done very often in machine learning, so pgmpy provides the direct methods `fit` and `predict` to simplify the whole process. Let's look at some code to understand how this works. To keep it simple, we will once again be working with the restaurant model, with each variable having two states.

```
# First let's import modules that we will be needing
In [1]: import numpy as np
In [2]: from pgmpy.models import BayesianModel
```

```
# Now let's create some random data over which we will train and
# test the model. Here we are creating 1000 data points with each
# value either 0 or 1.
In [3]: data = np.random.randint(low=0, high=2, size=(1000, 4))
In [4]: data
Out[4]:
array([[0, 1, 0, 0],
       [1, 1, 1, 0],
       [1, 1, 0, 0],
       ...,
       [1, 0, 0, 1],
       [1, 0, 1, 0],
       [1, 0, 0, 0]])
```

```
# Now in general machine learning problems it doesn't matter which
# column of the array represents which variable (until we use same
# order for both training and prediction) because all the values
# are on symmetrical axis but in graphical models each variable is
# different (in the way it is connected to other variables etc) so
# we will need to specify which columns of data are for which
# variable. For that we will use pandas.
```

```
In [5]: import pandas as pd
In [6]: data = pd.DataFrame(data, columns=['cost', 'quality',
                                           'location',
                                           'no_of_people'])
In [7]: data
Out[7]:
```

	cost	quality	location	no_of_people
0	0	1	0	0
1	1	1	1	0
2	1	1	0	0
3	0	1	1	1
4	1	1	1	0
5	1	0	1	0
6	0	0	0	0
7	0	0	1	0
..	...	...	...	...
993	0	0	1	1
994	0	0	0	0
995	0	0	0	0
996	1	0	0	0
997	1	0	0	1
998	1	0	1	0

```
999      1         0         0                   0
```

```
In [8]: train = data[:750]
```

```
# We will try to predict the no_of_people from our model. So for
# test data we will delete that column and then later on predict
# those values.
In [9]: test = data[750:].drop('no_of_people', axis=1)
In [10]: test
Out[10]:
      cost  quality  location
750     0        0         1
751     0        1         1
752     0        1         1
753     1        0         0
754     1        0         1
755     1        0         1
756     0        1         0
757     1        0         0
..    ...      ...       ...
992     0        0         0
993     0        0         1
994     0        0         0
995     0        0         0
996     1        0         0
997     1        0         0
998     1        0         1
999     1        0         0
```

```
# Now we will need to create the base network structure for the
# model.
In [11]: restaurant_model = BayesianModel(
                    [('location', 'cost'),
                     ('quality', 'cost'),
                     ('location', 'no_of_people'),
                     ('cost', 'no_of_people')])
In [12]: restaurant_model.fit(train)
```

```
# Fit computes the cpd of all the variables from the training data
# that we provided.
In [13]: restaurant_model.get_cpds()
Out[13]:
[<pgmpy.factors.CPD.TabularCPD at 0x7fc01c029be0>,
 <pgmpy.factors.CPD.TabularCPD at 0x7fc01c029eb8>,
```

```
<pgmpy.factors.CPD.TabularCPD at 0x7fc01c029e48>,
<pgmpy.factors.CPD.TabularCPD at 0x7fc01c029e80>]

# Now for predicting the values of no_of_people using this model
# we can simply call the predict method on our test data.
In [14]: restaurant_model.predict(test).values.ravel()
Out[14]:
array([[1, 1, 1, 1, 1, 0, 0, 0, 1, 0, 0, 0, 0, 1, 1, 0, 0, 1, 0,
  0, 0, 1, 1, 0, 1, 1, 1, 1, 1, 1, 0, 1, 1, 1, 1, 0, 1, 0,
  1, 0, 1, 0, 0, 0, 1, 1, 1, 1, 1, 1, 1, 0, 1, 1, 1, 1, 0,
  0, 1, 0, 1, 0, 0, 0, 1, 0, 1, 0, 1, 1, 0, 0, 0, 1, 1, 0,
  1, 1, 0, 1, 0, 1, 1, 0, 1, 1, 0, 1, 0, 1, 0, 0, 1, 0, 0,
  0, 1, 0, 1, 0, 0, 0, 0, 1, 1, 0, 1, 0, 0, 1, 0, 1, 1, 0,
  1, 0, 1, 1, 1, 0, 0, 1, 1, 0, 1, 0, 0, 0, 0, 1, 0, 0, 0,
  0, 1, 1, 0, 0, 0, 0, 0, 0, 0, 0, 0, 1, 1, 0, 0, 0, 0, 0,
  0, 0, 0, 1, 0, 1, 0, 0, 0, 0, 0, 1, 1, 1, 0, 0, 1, 0, 1,
  0, 1, 0, 1, 1, 1, 0, 0, 1, 1, 1, 0, 1, 1, 1, 0, 1, 1, 1,
  0, 0, 1, 1, 0, 1, 0, 1, 0, 1, 1, 0, 0, 0, 1, 1, 1, 1, 1,
  1, 0, 0, 1, 0, 1, 1, 0, 1, 0, 0, 1, 1, 1, 1, 1, 1, 0, 1,
  1, 0, 0, 0, 1, 1, 0, 0, 0, 0, 1, 1, 0, 0, 1, 0, 1, 1, 1,
  0, 0, 0]])
```

We can see here that using `fit` and `predict` has reduced a lot of work and simplified things. Also, in some cases, the training data we have might not represent the problem correctly. For example, let's say we know from prior knowledge that the probability of having a restaurant in a good location or a bad location is 0.5, but it is possible that the training set that we have has more data points for restaurants in good locations, which could eventually lead to bias in our model. In such cases, we could manually adjust the probability values in the CPDs so that they represent the actual problem correctly.

A comparison of variable elimination and belief propagation

In the previous sections, we saw that both belief propagation and variable elimination are inter-related. Belief propagation is an extension of the variable elimination algorithm on clique trees. So, one might think that they would have the same computational complexity. However, in reality, belief propagation has some advantages over variable elimination.

The major advantage is the ability to query over multiple variables of a model with a single computation (that is, calibration of the clique tree). Once the tree is calibrated, we could query about multiple variables without performing any further computation. However, in the case of variable elimination, we have to run the algorithm more than once. Thus, if we have such a problem, in which we need to query the model multiple times, we should definitely use belief propagation.

On the flipside, belief propagation also has a disadvantage over variable elimination. Clique trees are a memory-expensive data structure. Moreover, in belief propagation, we have to store the generated intermediate factors, whereas in the case of variable elimination, we just discard them. Belief propagation is also less flexible as compared to variable elimination, as the clique tree is fixed and predetermined. So, in the case of very huge networks, memory might become a constraint when using belief propagation.

In a nutshell, we can say that variable elimination is computationally expensive, whereas belief propagation is memory expensive. We have to consider the trade-offs to decide which algorithm to go for. If we have a very large network, then variable elimination would be an attractive solution as it wouldn't be expensive in terms of memory. However, in the case of smaller networks and multiple queries, where computational time matters, it would be better to go with the belief propagation approach.

Summary

In this chapter, we discussed two algorithms, namely variable elimination and belief propagation, to find the conditional probability and do MAP queries on the models. We also discussed how the elimination order of variables in variable elimination affects the running complexity of the algorithm. To select efficient ordering, we discussed a few algorithms. Then, we discussed MAP queries, using which we can approach our machine learning problems through graphical models. We also compared variable elimination and belief propagation and discussed the benefits of each of these and when to use them.

In the next chapter, we will discuss various algorithms for approximate inference, including sampling methods, using which we can do approximate inference over models. Approximate methods help us save computation when we don't need the computations to be exact.

4

Approximate Inference

In the previous chapter, we saw algorithms for exact inference on graphical models. The computational complexity of calculating exact inference is exponential to the tree width of the network. Hence, for much larger networks whose tree width is large, exact inference becomes infeasible. Also, in many of our real-life problems, we are not particularly concerned about the exact probabilities of random variables. Rather, we are much more interested in the relative probabilities of the states of variables. Therefore, in this chapter, we will discuss algorithms to perform approximate inference over networks. There are many algorithms for approximate inference, but the approach to find an approximate distribution remains the same in all of them. In most of these, we usually define a target class Q of easy distributions, and then from this class, we try to find the distribution that is closest to our actual distribution P_Φ and answer inference queries from this estimated distribution.

In this chapter, we will discuss:

- Approximate inference as an optimization problem
- Solving optimization problems using Lagrange multipliers
- Deriving a clique tree algorithm from an optimization problem
- The loopy belief propagation algorithm with code examples
- The expectation propagation algorithm with code examples
- The mean field algorithm with code examples
- The full particles and collapsed particles sampling methods
- The Markov chain Monte Carlo method

The optimization problem

Let's start with a little recap of exact inference. Assume that we have a factorized distribution in the following form:

$$P_{\Phi}(x) = \frac{1}{Z} \prod_{\phi \in \Phi} \phi(U_{\phi})$$

Here, Z is the partition function, ϕ are the factors in the network, and U_{ϕ} is the scope of the factor ϕ. In the case of exact inference, we computed $P_{\Phi}(x)$ and then answered queries over this distribution.

In the case of belief propagation, the end result of running the algorithm was a set of beliefs on the clusters and sepsets. This set of beliefs was able to represent the joint distribution $P_{\Phi}(X)$. So, in the case of exact inference, we tried to find a set of calibrated beliefs that was able to represent our joint distribution exactly. For approximate algorithms, we will try to select the set of beliefs from all the sets of beliefs that conform to the cluster tree and are best able to represent our original distribution $P_{\Phi}(X)$.

So now the question is, how do we compare the similarity between these two distributions? There are many methods that we can use to compute the relative similarity of the two distributions, for example, Euclidean distance, L_1 distance, and relative entropy. However, the problem with most of these methods is that we need to answer hard queries on $P_{\Phi}(x)$ to compute the distance, and the whole purpose of approximate inference is to avoid computing the exact joint distribution. By using relative entropy to measure the similarity between the distributions, we can avoid answering hard queries on $P_{\Phi}(X)$. Now, let's see how relative entropy is defined over distributions.

The relative entropy between two distributions P_1 and P_2 is defined as follows:

$$D(P_1 \| P_2) = E_{P_1}\left[\ln \frac{P_1(x)}{P_2(x)} \right]$$

The relative entropy is always non-negative and is 0 only when $P_1 = P_2$. Also, the relative entropy is a nonsymmetrical quantity, so $D(P_1 \| P_2) \neq D(P_2 \| P_1)$.

Now, in our case of approximate inference, we will use $D(Q \| P_{\Phi}(x))$ (not $D(P_{\Phi}(x) \| Q)$) because computing it also requires computing $P_{\Phi}(x)$). Then, we can find the value of Q, which minimizes $D(Q \| P_{\Phi}(x))$.

Summarizing our complete optimization problem, let's assume that we have a cluster tree T for a distribution P_Φ and are given following the set of beliefs:

$$Q = \left\{ \beta_i : i \in V_T \right\} \cup \left\{ \mu_{(i,j)} : (i,j) \in E_T \right\}$$

Here, C_i denotes the clusters in T, β_i denotes beliefs over C_i, and $\mu_{(i,j)}$ denotes beliefs over $Sep(i,j)$. This set of beliefs represents a distribution Q as follows:

$$Q = \frac{\prod_{i \in V_T} \beta_i}{\prod_{(i,j) \in E_T} \mu_{i,j}}$$

As the cluster tree is calibrated, it satisfies the marginal consistency constraints and therefore $\mu_{(i,j)}$ for each $(i,j) \in E_T$ are the marginals of β_i and β_j. Therefore, the set of calibrated beliefs Q satisfies the following equations:

$$Q(c_i) = \beta_i[c_i]$$

$$Q\left(s_{(i,j)}\right) = \mu_{i,j}\left[s_{i,j}\right]$$

Now, we can define our optimization problem by selecting Q from the space of calibrated sets **Q**:

$$Q = \left\{ \beta_i : i \in V_T \right\} \cup \left\{ \mu_{(i,j)} : (i,j) \cup E_T \right\}$$

This must be done such that it minimizes $D(Q \| P_\Phi)$ with the following constraints:

$$\mu_{i,j}\left[s_{i,j}\right] = \sum_{C_i - S_{i,j}} \beta_i(c_i) \; \forall (i,j) \in E_T \; \forall s_{i,j} \in Val\left(S_{i,j}\right)$$

$$= \sum_{c_i} \beta_i(c_i) = 1 \; \forall i \in V_T$$

To solve this optimization problem, we examine the different configurations of beliefs that satisfy the marginal consistency constraints and select the one that minimizes our objective entropy function $D(Q \| P_\Phi)$.

The energy function

In the previous section, we saw that to find the approximate distribution, we need to optimize the relative entropy $D(Q \| P_{\Phi})$, but computing the relative entropy requires us to compute a summation over all possible instantiations of χ. To avoid this, we will now try to transform our optimization function in the form of an energy function.

We know the following:

$$D(Q \| P_{\Phi}) = E_Q \left[\ln Q(\chi) \right] - E_Q \left[\ln P_{\Phi}(\chi) \right]$$

Using the product form of $P_{\Phi}(x)$, we have the following:

$$\ln P_{\Phi}(\chi) = \sum_{\phi \in \Phi} \ln \phi(U_{\phi}) - \ln Z$$

Also, we know that $H_Q(\chi) = -E_Q \left[\ln Q(\chi) \right]$. Using this in the preceding equation, we get the following:

$$D(Q \| P_{\Phi}) = H_Q(\chi) = -E_Q \left[\sum_{\phi \in \Phi} \ln(U_{\phi}) \right] + E_Q \ln Z$$

$$= -F \left[\tilde{P}_{\Phi}, Q \right] + \ln Z$$

Here, $F \left[\tilde{P}_{\psi}, Q \right]$ is the energy functional where:

$$F \left[\tilde{P}_{\Phi}, Q \right] = E_Q \left[\ln \tilde{P}(\chi) \right] + H_Q(\chi)$$
$$= \sum_{\phi \in \Phi} E_Q \ln \phi | + H_Q(\chi)$$

The important thing to note here is that Z in the relative entropy term doesn't depend on Q. Hence, minimizing the relative entropy $D(Q \| P_{\Phi})$ is equivalent to maximizing the energy function $F \left[\tilde{P}_{\Phi}, Q \right]$.

Now, the energy function has two terms. The first one is known as the **energy term**. The energy term is the summation of the expectations of the logarithm of the factors in ϕ. Therefore, in this term, each factor of ϕ appears separately. Hence, if these factors are small, then the expectations will be dealing with much fewer variables. The second term in the energy function is called the *entropy term* and it represents the entropy of Q. The complexity of computing this depends on our choice of Q.

Exact inference as an optimization

Before considering the approximate inference methods, let's solve the exact inference problem using the concepts that we have so far developed in this chapter. In the previous sections, we saw that maximizing the energy function is equivalent to minimizing the relative entropy between Q and $P_\Phi(x)$. So now, if we restrict ourselves to calibrated cluster trees, we can further simplify the objective function. Restricting ourselves to calibrated cluster trees allows us to rewrite the energy function in a factored form as a sum of terms, each depending directly on only one of the beliefs in Q. This form also reveals structure in the distribution, and is therefore a much better starting point for further analysis.

Given a cluster tree T with a set of beliefs Q and an assignment α, which maps factors in to clusters in T, we define the factored energy function as follows:

$$\tilde{F}\left[\tilde{P}_\Phi, Q\right] = \sum_{i \epsilon V_T} E_{C_i \, \beta_i}\left[ln \, \psi_i\right] + \sum_{i \epsilon V_T} H_{\beta_i}\left(C_i\right) - \sum_{(i,j)\epsilon T} H_{\mu_{i,j}}\left[S_{i-j}\right]$$

Here, ψ_i is the initial potential assigned to C_i:

$$\psi_i = \prod_{\phi, \alpha(\phi)=I} \phi$$

Here, $E_{C_i \, \beta_i}$ represents the expectation on the value C_i given the beliefs β_i.

The first term is a sum of terms of the form $E_{C_i \, \beta_i}\left[ln\psi_i\right]$. Here, ψ_i is a factor over the scope C_i and therefore, it maps from $Val(C_i)$ to $\mathbb{R}^+$. Hence, its logarithm is a function from $Val(C_i)$ to $\mathbb{R}^+$. The beliefs β_i are a distribution over $Val(C_i)$. We can therefore compute the expectation $\sum_{c_i} \beta_i(c_i) ln\psi_i$. The last two terms are the entropies of the beliefs associated with the clusters and sepsets in the tree. The important benefit of this reformulation is that all the terms are now local and hence represent a specific belief factor. We will see later on that this makes our tasks much simpler.

Now, using this form of the energy function, we can define the optimization problem. Now, as Q is factorized according to T, we can represent it with a set of calibrated beliefs. Marginal consistency is a constraint on the beliefs that requires neighboring beliefs to agree on the marginal distribution on their joint subset, which is equivalent to requiring that the beliefs be calibrated. Thus, we have the following constrained optimization problem:

$$Q = \left\{ \beta_i : i_T \right\} \left\{ \mu_{i,j} : \left(i - j \, \epsilon_T \right) \right\}$$

We want to optimize $F\left[\tilde{P}_\Phi, Q \right]$, where:

$$\mu_{i,j}\left[s_{i,j} \right] = \sum_{C_i, S_{i,j}} \beta_i\left(c_i \right) \quad \forall \left(i,j \right) \epsilon \, E_T, \forall s_{i,j} \, \epsilon \, Val\left(S_{i,j} \right)$$

$$\sum_{c_1} \beta_i\left(c_i \right) = 1 \quad \forall i \, \epsilon \, V_T$$

$$\beta_1\left(c_i \right) \geq 0 \quad \forall i_T, c_i \, \epsilon \, Val\left(C_i \right)$$

The constraints here are to ensure that the beliefs in Q are calibrated and represent legal distributions.

As we now have a constrained optimization problem, we can use the Lagrangian multipliers to solve this. Applying the Lagrangian multipliers, we get the following equation:

$$J = \tilde{F}\left[\tilde{P}_\Phi, Q \right]$$

$$- \sum_{i \epsilon V_T} \lambda_i \left(\sum_{c_i} \beta_i\left(c_i \right) - 1 \right)$$

$$- \sum_i \sum_{j \epsilon Nb_i} \sum_{s_{i,j}} \lambda_{j \to i}\left[s_{i,j} \right] \left(\sum_{c_i \, s_{i,j}} \beta_i\left(c_i \right) - \mu_{i,j}\left[s_{i,j} \right] \right)$$

Here, Nb_i is the neighbor of C_i in the clique tree. We have introduced Lagrange multipliers λ_i for each belief factor β_i to ensure that it sums up to 1. Also, for each pair of neighboring cliques i and j and their assignment to sepset $s_{i,j}$, we introduced a Lagrange multiplier $\lambda_{j \to i}\left[s_{i,j} \right]$ to ensure that the marginal distribution of $s_{i,j}$ in β_j is consistent with its values in the sepset beliefs $\mu_{(i,j)}$.

Now, we simply need to find the maximum value of the Lagrangian J and for that, we take its partial derivatives with respect to $\beta_i(c_i)$, $\mu_{i,j}[s_{i,j}]$ and the Lagrange multipliers:

$$\frac{\partial}{\partial \beta_i(c_i)} J = \ln \psi_i[c_i] \ln \beta_i(c_i) - 1 - \lambda_i - \sum_{j \in Nb_i} \lambda_{j \to i}[s_{i,j}]$$

$$\frac{\partial}{\partial \mu_{i,j\,s_{i,j}}} J = \ln \mu_{i,j}[s_{i,j}] + 1 + \lambda_{i \to j}[s_{i,j}] + \lambda_{j \to i}[s_{i,j}]$$

Now, equating these to 0 to find the maxima, we get the following equations:

$$\beta_i(c_i) = \exp\{-1 - \lambda_i\} \psi_i \mid c_i \prod_{j \in Nb_i} \exp\{-\lambda_{j \to i}[s_{i,j}]\}$$

$$\mu_{i,j}[s_{i,j}] = \exp\{-1\} \exp\{-\lambda_{i \to j}[s_{i,j}]\} \exp\{-\lambda_{j \to i[s_{i,j}]}\}$$

These equations describe beliefs as functions of terms of the form $\exp\{-\lambda_{i \to j}[s_{i,j}]\}$. These terms play the role of a message $\delta_{i \to j}$. To make this more explicit:

$$\delta_{i \to j}[s_{i,j}] = \exp\left\{-\lambda_{i \to j}[s_{i,j}] - \frac{1}{2}\right\}$$

Rewriting the equation, we get the following:

$$\beta_i(c_i) = \exp\left\{-\lambda_i - 1 + \frac{1}{2}|Nb_i|\right\} \psi_i(c_i) \prod_{j \in Nb_i} \delta_{j \to i}[s_{i,j}]$$

$$\mu_{i,j}[s_{i,j}] = \delta_{i \to j}[s_{i,j}] \delta_{j \to i}[s_{i,j}]$$

We can now rewrite the message $\delta_{i \to j}$ as follows:

$$
\begin{aligned}
\delta_{i \to j}\, s_{i,j} &= \frac{\mu_{i,j}\left[s_{i,j}\right]}{\delta_{j \to i}\left[s_{i,j}\right]} \\
&= \frac{\sum_{C_i\, S_{i,j}} \beta_i\left(C_i, s_{i,j}\right)}{\delta_{j \to i}\left[s_{i,j}\right]} \\
&= \exp\left\{-\lambda_i - 1 + \frac{1}{2}\left|Nb_i\right|\right\} \sum_{C_i - S_{i,j}} \psi_i\left(c_i\right) \prod_{k \in Nb_i - \{j\}} \delta_{k \to i}\left[s_{i,k}\right]
\end{aligned}
$$

Note that the term $\exp{-\lambda_i 1 + \frac{1}{2}\left|Nb_i\right|}$ is a constant as it doesn't depend on c_i. When we combine these equations, we can solve for λ_i to ensure that this constant normalizes the clique beliefs β_i.

The propagation-based approximation algorithm

The propagation-based approximation algorithm is a more generalized version of the belief propagation algorithm and works on the same principle of passing messages. In the case of exact inference, we used to construct a clique tree and then passed messages between the clusters. However, in the case of the propagation-based approximation algorithms, we will be performing message passing on cluster graphs.

Let's take the simple example of a network:

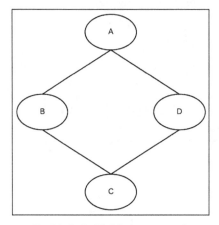

Fig 4.1: A simple Markov network

It is possible to construct multiple cluster graphs for this network. Let's take the example of the following two cluster graphs:

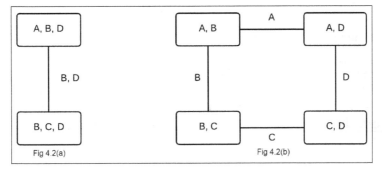

Fig 4.2: Cluster graphs for the network in Fig 4.1

Fig 4.2 shows two possible cluster graphs for the network in Fig 4.1. The cluster graph in Fig 4.2(a) is a clique tree and the clusters are *(A, B, C)* and *(B, C, D)*. Whereas, the cluster graph in Fig 4.2(b) has four clusters *(A, B)*, *(B, C)*, *(C, D)*, and *(A, D)*. It also has loops:

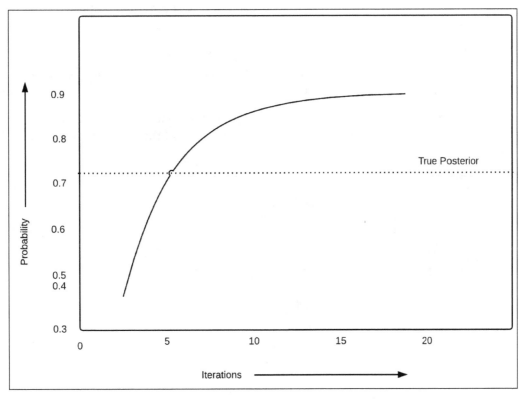

Fig 4.3: Change of estimated probability with a number of iterations

Let's assume that the factors for the network are such that it's more likely for the variables to agree with the same state than different states, that is, $\beta\left(a^0,b^0\right)$ and $\beta\left(a^1,b^1\right)$ are much larger than $\beta\left(a^0,b^1\right)$ and $\beta\left(a^1,b^0\right)$, and so on. Applying the message passing algorithm, messages will be passed from (A, B) to (B, C) to (C, D) to (A, D) and then again from (A, D) to (A, B). Also, let's consider that the strength of the message $\mu_{(A,B),(B,C)}$ increases the belief that $b = 0$. So now, when the cluster passes the message, it will increase the belief that $c = 0$ and so on. So finally, when the message reaches (A, B), it will increase the belief of A being 0, which in turn also increases the chances of B being 0. Hence, in each iteration, because of the loop in the network, the probability of A being 0 keeps on increasing until it reaches a convergence point, as shown in the graph in Fig 4.3.

Cluster graph belief propagation

In the case of exact inference, we had imposed two conditions on cluster graphs that led us to the clique trees. The first one was that the cluster graph must be a tree and should have no loops. The second condition was that it must satisfy the running intersection property. Now, in the case of the cluster graph belief propagation, we remove the first condition and redefine a more generalized running intersection property.

We say that a cluster graph satisfies a running intersection property if, whenever there is a variable X, and $X \in C_i$ and $X \in C_j$, there is only one path from C_i to C_j through which messages about X flow.

This new generalized running intersection property leaves us another question, "how do we define sepsets now?". Let's take the example of the following two cluster graphs in Fig 4.4:

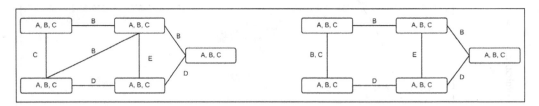

Fig 4.4: Two different clusters for the same network

In the case of exact inference, our sepsets used to be the common elements in the clusters. However, as we can see in the examples in the Fig 4.4, the same variable is common in multiple clusters. Therefore, to satisfy our running intersection property, we can't have it in the sepset of all the clusters.

In the case of clique trees, we performed inference by calibrating beliefs. Similarly, in the case of cluster graphs, we also say that the graph is calibrated if, for each edge (i, j) between the clusters C_i and C_j, we have the following:

$$\sum_{C_i - S_{i,j}} \beta_i = \sum_{C_j - S_{i,j}} \beta_i$$

Looking at the preceding equation, we can also say that a cluster graph is calibrated if the marginal of a variable X is same in all clusters containing X in their sepsets.

To analyze the computational benefits of this cluster graph algorithm, we can take the example of a grid-structured Markov network, as shown in Fig 4.5:

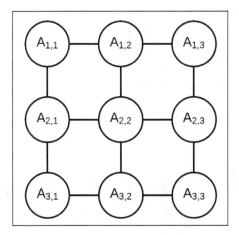

Fig 4.5: A 3 x 3 two-dimensional grid network

In the case of grid graphs, we are usually given the pair-wise parameters so they can be represented very compactly. If we want to do exact inference on this network, we would need separating sets that are as large as cutsets in the grid. Hence, the cost of doing exact inference would be exponential in n, where the size of the grid is $n \times n$. Whereas, if we are doing approximate inference, we can very easily create a generalized cluster graph that directly corresponds to the factors given in the network. We can see one such example in Fig 4.6:

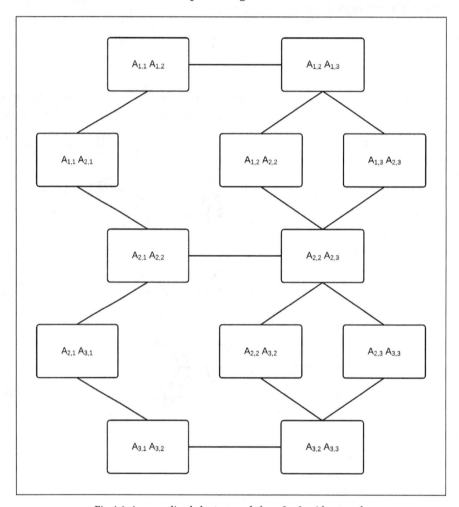

Fig 4.6: A generalized cluster graph for a 3 x 3 grid network

Each iteration of propagation in a cluster graph is quadratic in n.

Constructing cluster graphs

In our discussion so far, we have considered that we were already given the cluster graph. In the case of clique trees, we saw that different tree structures give the same result, but the computational cost varies in different structures. Also, in the case of cluster graphs, different structures have different computational costs, but the results also vary greatly. A cluster graph with a much lower computational cost may give very poor results compared to other cluster graphs with higher costs. Thus, while constructing cluster graphs, we need to consider the trade-off between computational cost and the accuracy of inference.

There are various approaches to construct cluster graphs. Let's discuss a few of them.

Pairwise Markov networks

In this class of networks, we are given potentials on each of the variables $\phi_i[X_i]$ and also pairwise potentials over some of the variables $\phi_{(i,j)}[X_i, X_j]$. These pairwise potentials correspond to the edges in the Markov network, and this kind of network occurs in many natural problems. For these kinds of networks, we add clusters for each of these potentials and then add edges between clusters with common variables. Taking the example of a 3 x 3 grid graph, we will have a cluster graph, as shown in Fig 4.7:

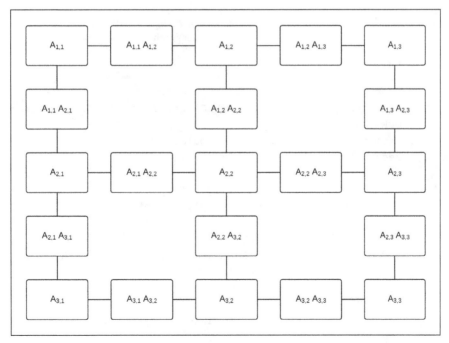

Fig 4.7: The cluster graph for a 3 x 3 grid when viewed as a pairwise MRF

Also, one thing worth noticing is that we can always reduce any network to a pairwise Markov structure and apply this transformation to construct cluster trees.

Bethe cluster graph

Pairwise Markov networks work well only for cases where we have factors with small scopes. However, in cases where the factors are complex, we won't be able to do the transformation of the pairwise Markov network. For these networks, we can use the Bethe cluster graph construction. In this method, we create a bipartite graph placing all the complex potentials on one side and the univariate potentials on the other side. Then, we connect each univariate potential with the cluster that has that variable in its scope, thus resulting in a bipartite graph, as shown in Fig 4.8:

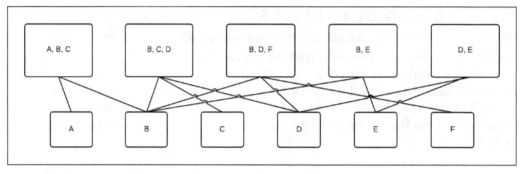

Fig 4.8: Cluster graph for a network over potentials {A, B, C}, {B, C, D}, {B, D, F}, {B, E}, and {D, E} viewed as a Bethe cluster graph

Its implementation with `pgmpy` is as follows:

```
In [1]: from pgmpy.models import BayesianModel
In [2]: from pgmpy.inference import ClusterBeliefPropagation as
        CBP
In [3]: from pgmpy.factors import TabularCPD
In [4]: restaurant_model = BayesianModel([
                        ('location', 'cost'),
                        ('quality', 'cost'),
                        ('location', 'no_of_people'),
                        ('cost', 'no_of_people')])
In [5]: cpd_location = TabularCPD('location', 2, [[0.6, 0.4]])
In [6]: cpd_quality = TabularCPD('quality', 3, [[0.3, 0.5, 0.2]])
In [7]: cpd_cost = TabularCPD('cost', 2,
                        [[0.8, 0.6, 0.1, 0.6, 0.6, 0.05],
                         [0.2, 0.1, 0.9, 0.4, 0.4, 0.95]],
                        ['location', 'quality'], [2, 3])
In  [8]: cpd_no_of_people = TabularCPD(
                        'no_of_people', 2,
```

```
                    [[0.6, 0.8, 0.1, 0.6],
                     [0.4, 0.2, 0.9, 0.4]],
                    ['cost', 'location'], [2, 2])
In   [9]: restaurant_model.add_cpds(cpd_location, cpd_quality,
                                     cpd_cost, cpd_no_of_people)
In  [10]: cluster_inference = CBP(restaurant_model)
In  [11]: cluster_inference.query(variables=['cost'])
In  [12]: cluster_inference.query(variables=['cost'],
                                  evidence={'no_of_people': 1,
                                            'quality':0})
```

Propagation with approximate messages

In the earlier section, we discussed a variant of belief propagation where we relaxed the constraint of having a clique tree, and did belief propagation on a cluster graph. In this section, we will take a different approach. Instead of relaxing on the structure, we will be approximating the messages passed between the clusters. Although this approach can be extended to work with cluster graphs as well, the scope of this book is only limited to clique trees.

Let's consider a simple pairwise Markov model, as shown in Fig 4.9. As discussed in the previous section, a pairwise Markov model is simply a Markov model with the factors $\phi_{i,j}$ associated with each edge $X_i - X_j$, along with the univariate factors ϕ_i corresponding to each random variable X_i. Thus, the following model will have factors such as $\phi_{A_{1,1},A_{1,2}}$, $\phi_{A_{1,1},A_{2,1}}$, and $\phi_{A_{1,1},A_{2,2}}$ along with $\phi_{A_{11}}$, $\phi_{A_{21}}$, $\phi_{A_{31}}$, and so on. Let's also assume that each random variable present in this network is binary.

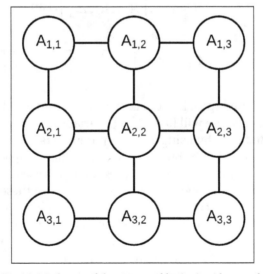

Fig 4.9: Markov model represented by 3 x 3 grid network

A cluster tree for this network can be created, as shown in Fig 4.10. Although this may not be an optimal cluster tree, it's a valid one as it satisfies the running intersection property, and each node represents a cluster of random variables present in the original network.

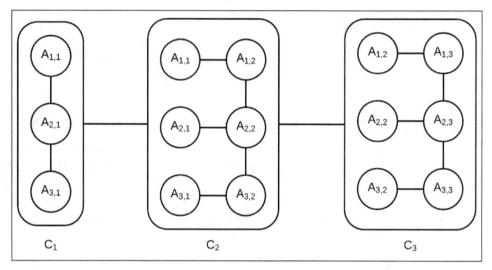

Fig 4.10: Cluster tree corresponding to the Markov model in Fig 4.9

In our previous discussion about the cluster tree (or clique tree), we never discussed the internal structure of each cluster, but the internal structure of the cluster becomes important in the context of this algorithm. For the calibration of the previously mentioned clique tree, we need to transmit message across the clusters. Suppose the message from C_1 to C_2, that is $\delta_{1 \to 2}\left[A_{1,1}, A_{2,1}, A_{3,1}\right]$, can be approximated by its factored form as follows:

$$\delta_{1 \to 2}\left[A_{1,1}, A_{2,1}, A_{3,1}\right] = \delta_{1 \to 2}\left[A_{1,1}\right]\delta_{1 \to 2}\left[A_{2,1}\right]\delta_{1 \to 2}\left[A_{3,1}\right]$$

We can see that the factored form is more compact as compared to the original message. The original message will have $2^3 = 8$ variables, whereas the factored form can be represented only by using $2 * 3 = 6$ parameters (two parameters for each variable as they are assumed to be binary). However, this compact representation helps us to save only two variables. So, the question that arises is whether the approximation is worth the savings or not. How can we use these approximations to compute the inference? We can get similar saving even if we just use some approximation that is richer than the naive independence assumption we used earlier. Even if we use approximations by exploiting the conditional independence among the random variables represented by the chain structure $A_{1,1} - A_{2,1} - A_{3,1}$, the question still remains the same: how can we use these approximations to compute the inference?

Before answering these questions, let's discuss factor sets. A factor set $\vec{\phi} = \{\phi_1,...,\phi_n\}$ provides a compact representation of $\phi_1 \cdot \phi_2 \cdots \phi_n$. Thus, the product of two factor sets is nothing but their union. For example, suppose $\vec{\phi}_1 = \{\phi_{11},...,\phi_{1n}\}$ and $\vec{\phi}_2 = \{\phi_{21},...,\phi_{2m}\}$. Then, their product should be $\phi_{11} \cdot \phi_{12} \cdots \phi_{1n} \cdot \phi_{21} \cdot \phi_{22} \cdots \phi_{2m}$, which can be written as a factor set of $\{\phi_{11},...,\phi_{1n}\} \cup \{\phi_{21},...,\phi_{2m}\}$.

Coming back to our previous question, how can we use these approximations to compute the inference? Let's assume that we somehow factorized the message from cluster C_1 to cluster C_2, that is $\delta_{1 \to 2}$ into a factor set $\vec{\delta}_{1 \to 2}$ consisting of univariate factors. Similarly, consider that we factorized the message from cluster C_3 to cluster C_2, and $\delta_{3 \to 2}$ into a factor set $\vec{\delta}_{3 \to 2}$ consisting only of univariate terms. To compute the belief of cluster $\vec{\delta}_{3 \to 2}$, we need to multiply the initial potential of C_2, that is C_2 with messages $\delta_{1 \to 2}$ and $\delta_{3 \to 2}$. As both the messages $\delta_{1 \to 2}$ and $\delta_{3 \to 2}$ have been factored into factor sets consisting only of univariate factors, the network structure of cluster C_2 remains unchanged (as shown in Fig 4.11). That is, no extra edge between any two variables is added as none of the factors from the message represent interaction among the random variables:

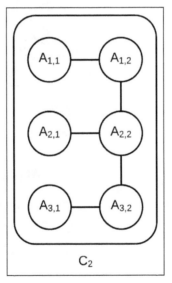

Fig 4.11: Internal network structure of cluster C_2 remains unchanged.
It is still a tree with tree width of two.

As the cluster C_2 has a tree structure internally, we can apply any exact inference algorithm to compute the marginals of the random variables present in this cluster.

If we use a richer approximation that exploits the chain structure of the cluster C_1 to compute the message $\delta_{1\to 2}$, it will contain factors representing interactions among A_{11}, A_{21} and A_{21}, A_{31}. When this message is multiplied with ψ_2 along with $\delta_{3\to 2}$, it will modify the network structure of C_2; it will introduce an edge between $A_{11} - A_{21}$ as well as an edge between $A_{21} - A_{31}$, as shown in Fig 4.12. Still, the network has a tree width of two and we can still use exact inference to compute the marginals of the random variables present in this cluster.

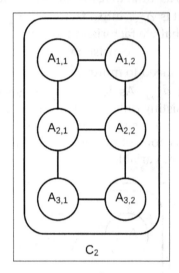

Fig 4.12: The internal network structure of C_2 with a richer approximation of $\delta_{1\to 2}$

So we can see how these approximations can help us in computing the inference.

Message creation

Now, the question is, how do we compute these messages, or more precisely, how do we factorize the message from cluster C_i to C_j, that is $\delta_{i\to j}$, into factor sets?

To answer this question, let's go back to the first principle method of computing a message from C_i to C_j. $\delta_{i\to j}$ is computed as follows:

$$\delta_{i\to j} = \sum_{C_i - S_{i,j}} \psi_i \prod_{neighbor(i)-\{j\}} \delta_{k\to i}$$

If all the messages from neighbors $\delta_{k \to i}$ are already factorized into factor sets, then their product is nothing but the union of their corresponding factor sets. The initial potential ψ_i can be factorized into a factor set of all the initial factors present in the cluster. The final factor product can be computed by the union of all the factor sets.

To compute the message, we also need to marginalize the after-product. To marginalize a factor set $\vec{\phi}$ with respect to a variable X, we need to couple all the factors containing X and marginalize them. So, like the product of a factor set, marginalizing it doesn't present any problems. So, the major problem lies in factorizing the marginal probabilities into a factor set. In a clique tree, the results from marginalizing a clique would not satisfy any conditional independence, so it can't be factorized into a factor set. However, for efficient inference, we want the messages to be factorized into a factor set. This can be achieved by approximating the message by a family of distributions that can be factorized. It turns out that there is a family of distributions that can be approximated for these messages and that the distribution is simply the product of the marginals of the individual variables present in the messages. The message is often not normalized, so it is not treated as a distribution. However, we can normalize the message and treat it as a distribution. To compute the marginals, we can use any of the exact inference algorithms that we discussed earlier, such as variable elimination or belief propagation.

Summarizing all these points, we can create an algorithm to compute the approximate messages to be transmitted between clusters in the clique tree:

1. Create a factor set $\vec{\phi}$ by the union of all the factor sets corresponding to the initial cluster potential as well as the input messages received.

2. Initialize an inference data structure $\mathcal{U}$ with this factor set to perform exact inference. It could be a clique tree in the case of belief propagation or a set of factors in the case of variable elimination.

3. Perform inference on $\mathcal{U}$ to compute the marginals of variables to be present in the final message.

4. The factor set of the marginals is the output message.

For example, let's try to work out how to create the messages $\delta_{1\rightarrow2}$ and $\delta_{2\rightarrow3}$ for the cluster tree represented in Fig 4.11, starting with $\delta_{1\rightarrow2}$. This can be computed by creating a factor set $\vec{\phi}$ as the union of $\vec{\psi}_1$ (factor set corresponding to ψ_1) and the input message. As there is no input message for this cluster, $\vec{\phi}$ will be $\{\phi A_{11}, A_{21}, \phi A_{21}, A_{31}, \phi A_{11}, \phi A_{21}, \phi A_{31}\}$. To compute the marginals for A_{11}, A_{21}, and A_{31} using the belief propagation method, we could use a clique tree, as shown in Fig 4.13:

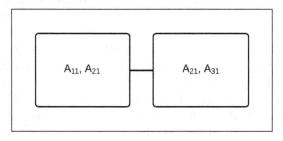

Fig 4.13: A Clique tree to compute the marginals of A_{11}, A_{21}, and A_{31}

The factor set representing the message from C_1 to C_2, that is $\delta_{1\rightarrow2}$, formed by the marginals of A_{11}, A_{21}, and A_{31} will be $\{\vec{\phi}A_{11}, \vec{\phi}A_{21}, \vec{\phi}A_{31}\}$.

Similarly, to compute $\delta_{2\rightarrow3}$, the first step is to create $\vec{\phi} = \delta_{1\rightarrow2} \cup \vec{\psi}_2$, where $\vec{\psi}_2$ represents the factor set corresponding to ψ_2. Then, we create an inference data structure for exact inference to compute the marginals of A_{12}, A_{22}, and A_{32} and initialize with $\vec{\phi}$. As $\delta_{1\rightarrow2}$ contains only univariate factors, the structure of C_2 remains unchanged. Fig 4.14 represents the clique tree that can be used as an inference data structure to compute the marginals of A_{12}, A_{22}, and A_{32}:

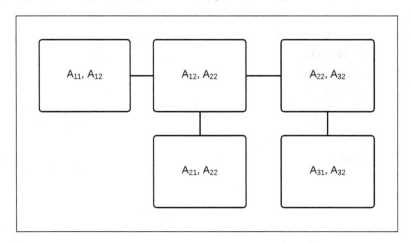

Fig 4.14: A clique tree to compute the marginals of A_{12}, A_{22}, and A_{32}

Inference with approximate messages

In the previous section, we discussed the methods of creating messages to transmit between clusters. Once we have these messages, the next task is to perform inference on the clique tree. While discussing exact inference, we discussed two methods of performing inference on a clique tree, one being the sum-product algorithm, the other being the sum-product-divide or belief update algorithm. For the exact inference, both of these algorithms will give the same result, but in the case of approximate inference, they are not the same.

Before we discuss these steps in detail, let's look at the difference between the exact and approximate inference algorithms. Once the tree is calibrated, the beliefs so computed don't represent the joint probability distribution of all the variables present in the cluster (as it was in the case of exact inference). So, to answer queries about the variables present in the cluster, we can't just marginalize other variables from the belief. Instead, after calibration, we have the factor sets of beliefs parameterizing the network structure of the corresponding cluster. In the previous example, after the clique tree is calibrated, the belief for the cluster C_2 can be factorized as follows:

$$\vec{\beta}_2 = \vec{\psi}_2 \cup \vec{\delta}_{1 \to 2} \cup \vec{\delta}_{3 \to 2}$$

The factors present in the factor set $\vec{\beta}_2$ parameterize the network structure of cluster C_2. As the network structure allows tractable inference, we can answer queries about these variables using inference methods such as variable elimination or belief propagation.

Sum-product expectation propagation

The sum-product expectation propagation algorithm is similar to the sum-product algorithm we discussed for exact inference, except that we modify the procedure to compute the message. There, we computed the message by summing out (or marginalizing) the variable from the product of factors. Here, we compute the message as discussed in the previous section. Similar to the exact inference equivalent, in the case of approximate inference for calibration of the clique tree, we require two passes, one upward and one downward. So, unlike the previous approximate inference, it converges in a fixed number of steps.

Let's start with a simple example, as shown in Fig 4.15:

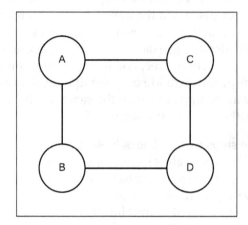

Fig 4.15: Simple pairwise Markov network consisting of four random variables

Suppose the factors associated with the given Markov model are as follows:

A	B	$\phi_1(A,B)$
a_0	b_0	10
a_0	b_1	0.1
a_1	b_0	0.1
a_1	b_1	10

A	C	$\phi_2(A,C)$
a_0	c_0	5
a_0	c_1	0.2
a_1	c_0	0.2
a_1	c_1	5

C	D	$\phi_4(D,B)$
c_0	d_0	0.5
c_0	d_1	1
c_1	d_0	20
c_1	d_1	2.5

D	B	a_0
d_0	b_0	5
d_0	b_1	0.2
d_1	b_0	0.2
d_1	b_1	5

From the preceding factors, we can see that there is a strong coupling between the variables A and B. It seems that $A = B$. The potentials $\phi_1(A,C)$ and $\phi_4(D,B)$ indicate weaker coupling between A and C, and B and D.

If we perform the exact inference in this network, we find the following marginal posteriors:

$$P(a_0,b_0)=0.274 \quad P(c_0,d_0)=0.102$$

$$P(a_0,b_1)=0.002 \quad P(c_0,d_1)=0.018$$

$$P(a_1,b_0)=0.041 \quad P(c_1,d_0)=0.368$$

$$P(a_1,b_1)=0.682 \quad P(c_1,d_1)=0.512$$

Let's try to compute the marginals using the approximate inference method that we discussed now using pgmpy. The clique tree constructed is shown in Fig 4.16:

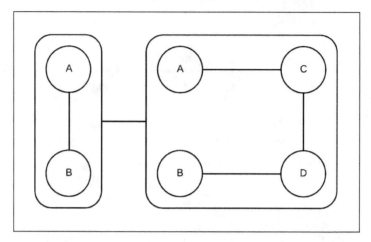

Fig 4.16: The clique tree constructed for the Markov model represented in Fig 4.18

```
In [1]: from pgmpy.factors import Factor
In [2]: from pgmpy.factors import FactorSet
In [3]: from pgmpy.models import MarkovModel
In [4]: from pgmpy.inference import VariableElimination
In [5]: from pgmpy.inference import BeliefPropagation
In [6]: import functools
In [7]: def compute_message(cluster_1, cluster_2,
                    inference_data_structure=
                        VariableElimination):
    """
    Computes the message from cluster_1 to cluster_2.
The messages are computed by projecting a factor set to
produce a set of marginals over a given set of scopes. The
factor set is nothing but the factors present in the models.

    The algorithm for computing messages between any two clusters
is:
* Build an inference data structure with all the factors
  represented in the cluster.
* Perform inference on the cluster using the inference data
  structure to compute the marginals of the variables present
  in the sepset between these two clusters.
    * The output message is the factor set of all the computed
      marginals.

    Parameters
    ----------
```

```
cluster_1: MarkovModel, BayesianModel, or any pgmpy supported
          graphical model
              The cluster producing the message
cluster_2: MarkovModel, BayesianModel, or any pgmpy supported
          graphical model
              The cluster receiving the message


   inference_data_structure: Inference class such as
                            VariableElimination or BeliefPropagation
              The inference data structure used to produce factor
              set of marginals
"""
# Sepset variables between the two clusters
sepset_var = set(cluster_1.nodes()).intersection(
                                    cluster_2.nodes())

# Initialize the inference data structure
inference = inference_data_structure(cluster_1)

# Perform inference
query = inference.query(list(sepset_var))

# The factor set of all the computed messages is the output
# message query would be a dictionary with key as the variable
# and value as the corresponding marginal thus the values
# would represent the factor set
   return FactorSet(*query.values())


In [8]: def compute_belief(cluster, *input_factored_messages):
       """
       Computes the belief a particular cluster given the cluster
       and input messages

       \delta_{j \rightarrow i} where j are all the neighbors of
       cluster i. The cluster belief is computed as:
    .. math::
       \beta_i = \psi_i \prod_{j \in Nb_i} \delta_{j \rightarrow i}

    where \psi_i is the cluster potential. As the cluster belief
    represents the probability and it should be normalized to sum
    up to 1.

    Parameters
    ----------
    cluster: MarkovModel, BayesianModel, or any pgmpy supported
             graphical model
              The cluster whose cluster potential is going to be
```

```
            computed.
    *input_factored_messages: FactorSet or a group of FactorSets
            All the input messages to the clusters. They should be
            factor sets

    Returns
    -------
    cluster_belief: Factor
        The cluster belief of the corresponding cluster
    """
    messages_prod = functools.reduce(lambda x, y: x * y,
                            input_factored_messages)

    # As messages_prod would be a factor set, so its corresponding
    # factor would be product of all the factors present in the
    # factorset
    messages_prod_factor = functools.reduce(lambda x, y: x * y,
                            messages_prod.factors)

    # Computing cluster potential psi
    psi = functools.reduce(lambda x, y: x * y,
                            cluster.get_factors())

    # As psi represents the probability it should be normalized
    psi.normalize()

    # Computing the cluster belief according the formula stated
    # above
    cluster_belief = psi * messages_prod_factor

    # As cluster belief represents a probability distribution in
    # this case, thus it should be normalized
    cluster_belief.normalize()

    return cluster_belief

In [9]: phi_a_b = Factor(['a', 'b'], [2, 2], [10, 0.1, 0.1, 10])
In [10]: phi_a_c = Factor(['a', 'c'], [2, 2], [5, 0.2, 0.2, 5])
In [11]: phi_c_d = Factor(['c', 'd'], [2, 2], [0.5, 1, 20, 2.5])
In [12]: phi_d_b = Factor(['d', 'b'], [2, 2], [5, 0.2, 0.2, 5])

# Cluster 1 is a MarkovModel A--B
In [13]: cluster_1 = MarkovModel([('a', 'b')])

# Adding factors
In [14]: cluster_1.add_factors(phi_a_b)
```

```
# Cluster 2 is a MarkovModel A--C--D--B
In [15]: cluster_2 = MarkovModel([('a', 'c'), ('c', 'd'),
                                  ('d', 'b')])

# Adding factors
In [16]: cluster_2.add_factors(phi_a_c, phi_c_d, phi_d_b)

# Message passed from cluster 1 -> 2 should the M-Projection of psi1
# as the sepset of cluster 1 and 2 is A, B thus there is no need to
# marginalize psi1
In [17]: delta_1_2 = compute_message(cluster_1, cluster_2)

# If we want to use any other inference data structure we can pass
# them as an input argument such as: delta_1_2 =
# compute_message(cluster_1, cluster_2, BeliefPropagation)
In [18]: beta_2 = compute_belief(cluster_2, delta_1_2)
In [19]: print(beta_2.marginalize(['a', 'b'], inplace=False))
```

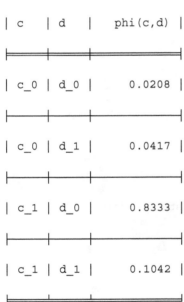

```
# Lets compute the belief of cluster1, first we need to compute the
# output message from cluster 2 to cluster 1
In [20]: delta_2_1 = compute_message(cluster_2, cluster_1)

# Lets see the distribution of both of these variables in the
# computed message
```

```
In [21]: for phi in delta_2_1.factors:
              print(phi)
```

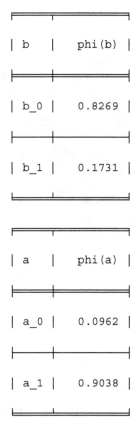

```
# The belief of cluster1 would be
In [22]: beta_1 = compute_belief(cluster_1, delta_2_1)
In [23]: print(beta_1)
```

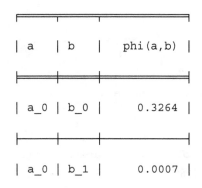

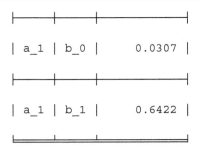

```
|   |   |   |             |
├───┼───┼───┼─────────────┤

| a_1 | b_0 |     0.0307 |


├───┼───┼───┼─────────────┤

| a_1 | b_1 |     0.6422 |

└───┴───┴───┴─────────────┘
```

Let's start with $\vec{\delta}_{1\to2}$. It can be computed by marginalizing ψ_1 with respect to A and B. Normalizing the messages to treat it as a distribution, we get $\phi_A(a_0) = 0.5$ and $\phi_A(a_1) = 0.5$. Similarly, for B we get $\phi_B(b_0) = 0.5$, $\phi_B(b_1) = 0.5$. Thus, $\vec{\delta}_{1\to2} = \{\phi_A, \phi_B\}$ or to put $\delta_{1\to2}$ would be as follows:

$$\delta_{1\to2}(a_0, b_0) = 0.25$$

$$\delta_{1\to2}(a_0, b_1) = 0.25$$

$$\delta_{1\to2}(a_1, b_0) = 0.25$$

$$\delta_{1\to2}(a_1, b_1) = 0.25$$

However, from exact inference we know the following:

$$\delta_{1\to2}(a_0, b_0) = 0.495$$

$$\delta_{1\to2}(a_0, b_1) = 0.005$$

$$\delta_{1\to2}(a_1, b_0) = 0.005$$

$$\delta_{1\to2}(a_1, b_1) = 0.495$$

We see that the approximate message loses the coupling between A and B. Thus, it is a poor approximation of the exact message. The problem with this approach is that the approximation of the message is done considering the impact of this message on the downstream cluster.

Similarly, if we compute $\delta_{2\to1}$, we get $\delta_{2\to1}(a_1) = 0.904$ and $\delta_{2\to1}(b_1) = 0.173$. This is again in contrast with the factor $\phi_1(A, B)$, which strongly suggests that $A = B$. When we combine the message with ψ_1, we get the belief for the cluster $\beta_1(a_0, b_1) = 0.001$ as follows:

$$\beta_1(a_0, b_0) = 0.326$$

$$\beta_1(a_0, b_1) = 0.001$$

$$\beta_1(a_1, b_0) = 0.031$$

$$\beta_1(a_1, b_1) = 0.642$$

This is fairly close to what the exact marginal suggests.

Belief update propagation

As we have seen in the previous example, when we had the message $\delta_{2\to1}$, we computed the posterior probability of A and B fairly close to the exact value. This raises the question, can we use the newly computed posterior probability to correctly approximate the message $\delta_{1\to2}$? The answer is, no, we can't. The reason for this is that, if we use the information that we got from C_j and use it to correct $\delta_{1\to2}$, we will be double-counting evidence. So, is there a way to get away with this double-counting yet still use the information?

If you recall, in the previous chapter, we discussed the belief update method that we used to compute the message from the cluster C_i to the cluster C_j as follows:

$$\delta_{i\to j} = \frac{\sum_{C_i - S_{i,j}} \beta_i}{\delta_{j\to i}}$$

So, using the belief update propagation, we can use the information $\delta_{2\to1}$ to modify $\delta_{1\to2}$. Let's see how to use it in the case of factor sets. The preceding equation can be translated for the factor set as follows:

$$\vec{\sigma}_{i\to j} = Approx(\beta_i)$$

$$\vec{\delta}_{i\to j} = \frac{\vec{\sigma}_{i\to j}}{\vec{\delta}_{j\to i}}$$

Here, $Approx(\beta_i)$ approximates the belief of the cluster C_i by a family of distributions that can be factorized. This is similar to what we did in the case of approximating a message by a family of factorized distributions. Let's go back to the example again to see how to implement it.

First, initialize all the messages to 1. In the first iteration, the value of $\vec{\delta}_{1\to2}$ is the same as what we computed earlier as $\vec{\delta}_{2\to1}$ is 1 and so would be $\vec{\delta}_{2\to1}$. In the second iteration to compute $\vec{\delta}_{1\to2}$, we will be using the value of β_1. Marginalizing β_1 with respect to B, we get $\phi_A(a_0)$ equals $0.326 + 0.001 = 0.327$ and $\phi_A(a_1)$ equals $0.031 + 0.642 = 0.673$. Similarly, marginalizing β_1 with respect to B, we get $\phi_B(b_0) = 0.326 + 0.031 = 0.357$ and $\phi_B(b_1) = 0.642 + 0.001 = 0.643$. So, $\sigma_{1\to2}$ will be as follows:

$$\sigma_{1\to2}(a_0, b_0) = 0.327 \times 0.357 = 0.116$$

$$\sigma_{1\to2}(a_0, b_1) = 0.327 \times 0.643 = 0.210$$

$$\sigma_{1\to2}(a_1, b_0) = 0.673 \times 0.357 = 0.240$$

$$\sigma_{1\to2}(a_1, b_1) = 0.673 \times 0.643 = 0.432$$

$\delta_{1\to2}$ can be computed by dividing $\sigma_{1\to2}$ with $\delta_{2\to1}$. Finally, we have to normalize $\delta_{1\to2}$ to treat it as a distribution.

The newly formed $\delta_{1\to2}$ can be viewed as a correction for β_2 in the next iteration, and so, it will be $\delta_{2\to1}$ for β_1. So, unlike the previous method, it doesn't converge in two steps; rather it requires multiple iterations of message passing between the two clusters, each correcting the other.

MAP inference

In the previous chapter, we studied MAP inference using variable elimination and max-product message passing in clique trees. In a similar fashion, we can apply max-product message passing on the cluster graph.

Recall that in the case of clique trees, the max-product message passing was analogous to their sum-product message passing algorithm, differing only in the way the message was computed. We used the maximization operation instead of summation. Also, in the case of cluster trees, the max-product message passing is analogous to their sum-product counterpart, maximizing the variable instead of summing it out. Unlike their sum-product counterpart, there is no guarantee of the convergence of this algorithm; it is more susceptible to nonconvergence. One reason for this is that the summation averages the messages, whereas maximization doesn't. Thus, it can't reduce oscillations.

Before going into further discussion about max-product message passing in cluster trees, let's discuss local optimality and decoding. We say that an assignment $\xi*$ has the local optimality property, if for each clique C_i in a max-calibrated clique tree, we have the following:

$$\xi*(C_i) \in \arg\max_{c_i} \beta_i(C_i)$$

The assignment to C_i in $\xi*$ optimizes the belief of C_i (that is β_i). The task of finding a locally optimal assignment $\xi*$, given a max-calibrated set of beliefs is known as **decoding**.

Just like the sum-product message passing on cluster trees, the max-product message passing will not give the exact max-marginal even after max-calibration. The beliefs so formed after max-calibration are called **pseudo max-marginals**.

Once we have the pseudo max-marginals by max-product message passing, we are left with the task of decoding these marginals. As discussed earlier, the task of decoding is nothing but finding a locally optimal assignment, and unlike clique trees, such assignments do not necessarily exist in the case of cluster graphs. Let's look at a simple example. Consider the cluster graph shown in Fig 4.17:

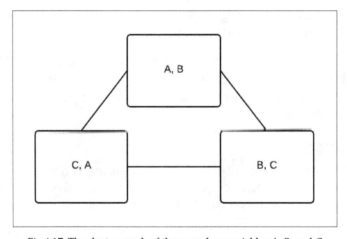

Fig 4.17: The cluster graph of three random variables *A*, *B*, and *C*

The beliefs after max-calibration are as follows:

A	B	$\beta_1(A,B)$
a_0	b_0	1
a_0	b_1	2
a_1	b_0	2
a_1	b_1	1

B	C	$\beta_2(B,C)$
b_0	c_0	1
b_0	c_1	2
b_1	c_0	2
b_1	c_1	1

A	C	$\beta_3(A,C)$
a_0	c_0	1
a_0	c_1	2
a_1	c_0	2
a_1	c_1	1

For example, to maximize $\beta_1(A,B)$, we can select the value of a_1, b_0. Thus, to maximize the belief $\beta_2(B,C)$, we have to select c_0. Now, we can see that the assignments a_1 and c_1 do not correspond to the maximum value of belief $\beta_3(A,C)$. No matter which assignment we choose, we can't obtain a single joint assignment that maximizes all three beliefs. These kinds of loops are called frustrating loops.

From the preceding example, we can create a simple hypothesis that if all the node beliefs are ambiguous, then there is no locally optimal joint assignment, but this is not always true. Let's take the example of the following beliefs:

A	B	$\beta_1(A,B)$
a_0	b_0	2
a_0	b_1	1
a_1	b_0	1
a_1	b_1	2

B	C	$\beta_2(B,C)$
b_0	c_0	2
b_0	c_1	1
b_1	c_0	1
b_1	c_1	2

A	C	$\beta_3(A,C)$
a_0	c_0	2
a_0	c_1	1
a_1	c_0	1
a_1	c_1	2

We can see that assignments a_0, b_0, and c_0, as well as a_1, b_1, and c_1 are locally optimal.

We saw some cases where there are no locally optimal assignments and there are cases where we can find locally optimal assignments. So, the basic question that arises is, how do we find locally optimal assignments, if any exist?

From the definition of local optimality, we can say that an assignment is locally optimal if and only if it selects optimal assignments from each cluster. Keeping this in mind, we can now assign labels to each assignment in a cluster. The label of the assignment can be "legal", if it optimizes the belief of that cluster, or "illegal" if it doesn't. So now, the decoding task is converted into a task of finding an assignment such that it is the legal value for all the clusters. This is nothing but a constraint satisfaction problem, where the constraints are obtained from the local optimality. The detailed survey of the constrained satisfaction problem is beyond the scope of this book. Thus, given a max-product calibrated cluster graph, we can convert it into a **constrained satisfaction problem (CSP)** simply by taking the belief of each cluster, and changing each assignment that locally optimizes the belief to 1 and the rest to 0. We then run a CSP solution method. If the outcome is an assignment that achieves 1 in every cluster belief, then the assignment is guaranteed to be a locally optimal assignment. For example, one of the CSP solution methods can be defined in terms of the Markov network, where all the entries are either 1 for legal assignments or 0 for illegal ones. Thus, CSP is simply finding the MAP assignment in a Markov model with {0, 1} valued beliefs. The CSP problem is itself an NP-hard problem. Thus, we can't guarantee that we would be able to find a locally optimal assignment efficiently, even if it existed.

The following is the implementation using pgmpy:

```
In [1]: from pgmpy.models import BayesianModel
In [2]: from pgmpy.factors import TabularCPD
In [3]: from pgmpy.inference import ClusterBeliefPropagation as
        CBP

# Create a bayesian model as we did in the previous chapters
In [4]: model = BayesianModel([
                    ('rain', 'traffic_jam'),
                    ('accident', 'traffic_jam'),
                    ('traffic_jam', 'long_queues'),
                    ('traffic_jam', 'late_for_school'),
                    ('getting_up_late', 'late_for_school')])
In [5]: cpd_rain = TabularCPD('rain', 2, [[0.4], [0.6]])
In [6]: cpd_accident = TabularCPD('accident', 2, [[0.2], [0.8]])
In [7]: cpd_traffic_jam = TabularCPD(
                    'traffic_jam', 2,
                    [[0.9, 0.6, 0.7, 0.1],
                     [0.1, 0.4, 0.3, 0.9]],
                    evidence=['rain', 'accident'],
                    evidence_card=[2, 2])
```

```
In [8]: cpd_getting_up_late = TabularCPD('getting_up_late', 2,
                                [[0.6], [0.4]])
In [9]: cpd_late_for_school = TabularCPD(
                       'late_for_school', 2,
                       [[0.9, 0.45, 0.8, 0.1],
                        [0.1, 0.55, 0.2, 0.9]],
                       evidence = ['getting_up_late',
                                'traffic_jam'],
                       evidence_card=[2, 2])
In [10]: cpd_long_queues = TabularCPD('long_queues', 2,
                                [[0.9, 0.2],
                                 [0.1, 0.8]],
                                evidence=['traffic_jam'],
                                evidence_card=[2])
In [11]: model.add_cpds(cpd_rain, cpd_accident, cpd_traffic_jam,
                   cpd_getting_up_late, cpd_late_for_school,
                   cpd_long_queues)
In [12]: cbp_inference = CBP(model)
In [13]: cbp_inference.map_query(variables=['traffic_jam',
                                    'late_for_school'])
In [14]: cbp_inference.map_query(variables=['traffic_jam'],
                        evidence={'accident': 1,
                                'long_queues': 0})
```

Sampling-based approximate methods

In the previous sections, we discussed a class of approximate methods that used factor manipulation methods to answer approximate queries on the models. Now, in this section, we will be discussing a very different approach to approximate inference. In this method, we will try to estimate the original distribution by instantiating all the variables or a few variables of the network. Using these instantiations, we will try to answer queries on the model. The methods using instantiations are generally known as **particle-based methods**, and each instantiation is known as a **particle**.

There are many variations of the way we select particles or create instantiations of the variables. For example, we can either create particles using a deterministic process, or we can sample particles from some distribution. Also, we can have different notions of a particle. For example, we can have a full assignment of all the variables in the network, commonly known as **full particles**, or we can have assignments only to a subset $P(x|w)$ of variables of the network representing the conditional probability $P(x|w)$. These are commonly known as **collapsed particles**. The main problem with full particles is that each particle is able to represent only a very small part of the whole space, and therefore, for a reasonable representation of the distribution, we need many more particles than are needed for collapsed particles.

In general, in the case of sampling methods, to approximate the values of queries, we generate some particles, and then, using these particles, we try to estimate the value or the expectation of the query relative to each of the generated particles and aggregate these to get the final result.

Also, concepts such as forward sampling and likelihood weighting, discussed in the next sections, only apply to Bayesian networks and not to Markov networks.

Forward sampling

The simplest approach to the generation of particles is **forward sampling**. In the case of forward sampling, we generate random samples $\xi[1]$, $\xi[2]$, ..., $\xi[M]$ from the distribution $P(X)$.

Let's take the example of generating particles using our restaurant model:

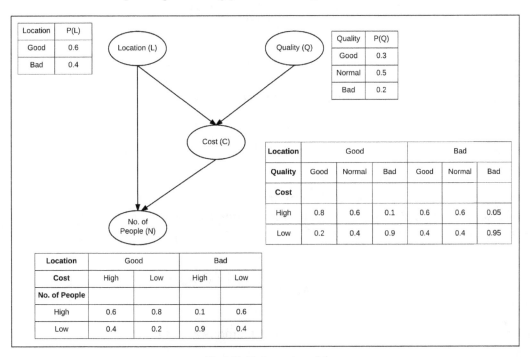

Fig 4.18: Restaurant model

We start by simply selecting a state of the variable *Location* with the probabilities *0.6* and *0.4*. Let's say we select the location of the restaurant to be good and select the quality to be good as well. Now, knowing the observed states of *Location* and *Quality*, we can now select the state of *Cost* to be high with the probability *0.8* and low with the probability *0.2*. Similarly, selecting a state for *No. of People*, we will get a single full particle over our restaurant model. To generate *M* particles, we repeat the same process *M* times to get M instantiations of the variables.

The main thing to notice is that we start with sampling variables that have no parents, and do it in an order such that before we sample any variable, we already have the values for all the parents of that variable.

After generating some particles, we can estimate the expectation of some function f using these particles as follows:

$$\hat{E}_D(f) = \frac{1}{M} \sum_{m=1}^{M} f(\xi \mid m)$$

Now, for a case when we want to compute the probability of some event $Y = y$, using these particles, we can simply calculate the fraction of particles satisfying the event. So, we can compute the probability $P(Y = y)$ as follows:

$$\hat{P}_D(y) = \frac{1}{M} \sum_{m=1}^{M} 1\{y[m] = y\}$$

So, taking the example of our restaurant model and computing the probability of *Cost* to be high, $P(C^0)$ of the restaurant is equivalent to getting the fraction of particles satisfying these values:

$$P(C^0) = \frac{1}{M} \sum_{m=1}^{M} 1\{\xi[m] < C >= 0\}$$

The key points to note in the case of particle methods are as follows:

- The result of the inference using sampling highly depends on the number of particles that we used for inference. It is quite possible that generating a very small number of samples will not represent our original distribution at all, and will thus give very inaccurate results.
- We can use the same particles to answer multiple queries, and therefore, sampling methods are very effective when we need to query the model multiple times.

Conditional probability distribution

Until now, we have only discussed computing the marginal probability of the form $P(Y = y)$ over variables, but in the real world, we are mostly working with conditional probability distributions rather than marginal distributions. Now, with sampling methods, we have multiple ways of approaching the problem of conditional distributions, but all of them turn out to be significantly harder than computing marginals.

Let's say we want to compute the probability of $P(y | E = e)$. The first approach that we can think of is to generate particles normally from the distribution and then reject the samples that don't satisfy the condition $E = e$. This method is known as rejection sampling. However, with this method, we will be wasting a lot of particles and thus increasing the computational cost. The real problem arises when the probability values of these events are very low. So, let's say $P(E = e) = 0.005$ and we generate 10,000 samples. Then we will have only around 50 samples that will satisfy our conditions. In general, to generate $M*$ samples that satisfy our conditions, we will have to generate $M = M* / P(e)$ samples. So, this method turns out to be really expensive when the probabilities of the variables are very small.

Another approach that we can take is to have separate estimators for $P(e)$ and $P(y, e)$, and after computing these, we can easily compute $P(y|e)$ using the chain rule. Also, with proper bounds on $P(e)$ and $P(y, e)$, we can get a good approximation for $P(y|e)$. The problem, however, is that to get a low relative error on $P(e)$, we will need samples that again depend on the value of $P(e)$. Also, a good bound on the relative error of $P(e)$ doesn't guarantee any bound on $\dfrac{P(y,e)}{P(e)}$. So, once again, we are stuck with the same problem that we saw earlier.

Likelihood weighting and importance sampling

In the previous section, we saw that the rejection method was very expensive because we were generating particles that were not consistent with our evidence, and then ultimately rejecting them. So, one possible solution to this problem is to generate particles that are more relevant to our event. We will be exploring this solution in this section.

To make the samples relevant to our evidence, we can force the sampling method to only take those values that we have observed. Taking the example of our restaurant model, let's say that we have observed that the location is good. Then, every time when generating a sample, we will only select the *Location* variable to be good. In this way, we can have observations that are consistent with our evidence, but this method leads to another problem. Let's say that we have observed that *Cost* is high, so when we generate samples forcing *Cost* to be high, we will still have the probability of *Location* to be good as being *0.6*, whereas as we have observed the cost to be high, it should have increased. The reason why this is happening is that this method fails to take into account the fact that the probability of the cost to be high is greater when the location is good than when the location is bad. To account for this error, we introduce weighting terms with each particle, which is equal to the probability of getting the forced state, given the states of other variables. Therefore, for the sample $L = good$, $Q = good$, $C = high$, and $N = low$, we will have the weighting 0.8 because $= 0.8$. Now, if multiple variables are forced, say $C = good$ and $N = high$, then for the sample, $L = good$, $Q = good$, $C = good$, and $N = high$, we will have the weighting *0.6 X 0.8 = 0.48*.

Generalizing this whole concept of assigning weighting to particles results in an algorithm known as **likelihood weighting**. Using this algorithm, we generate *weighted particles*. Using this likelihood weighting algorithm, we compute a set of M samples and their weights $P(C = good \mid L = good, Q = good)$, and using these samples, we can now estimate the conditional distribution $D = <\xi[1], w[1]>, <\xi[2], w[2]>, <\xi[M], w[M]>$ as follows:

$$\hat{P}_D(y \mid e) = \frac{\sum_{m=1}^{M} w[m] 1\{y[m] = y\}}{\sum_{m=1}^{M} w[m]}$$

Looking closely, we can see that this method is a generalization of forward sampling. In the case of forward sampling, each of the particles had the weight as 1, therefore, the numerator was simply the total number of particles satisfying the event and the denominator was the total number of particles. Also, it's very important to note that in the case of forward sampling, we can use these weighted particles to compute the probability of any event.

Importance sampling

As it turns out, likelihood weighting is a special case of a more generic method known as *importance sampling*. In this section, we will talk about importance sampling and show how likelihood weighting is derived from it.

Importance sampling is an approach used to estimate the expectation of a function $f(x)$ relative to some distribution $P(X)$, known as **target distribution**. As we saw in the previous sections, we can easily do this by generating particles $\xi[1], \xi[2], , \xi[M]$ from P and then estimating the following:

$$E_P[f] \approx \frac{1}{M} \sum_{m=1}^{M} f(x[m])$$

However, in some cases, we may want to generate samples from some other distribution Q, known as *proposal distribution* or *sampling distribution*, for whatever reason (for instance, it might be impossible or computationally very expensive to generate samples from P). For example, P might be a posterior distribution of a Bayesian network and hence, computing it may be very expensive. To deal with such problems, in this section, we will discuss methods to get expectation estimates relative to the distribution P, by generating samples from some other distribution Q.

So now, if we are generating our samples from the distribution Q, we can't simply use it to compute our expectation value. We need to adjust our estimator to compensate for this incorrect sampling. One obvious way to do this is as follows:

$$E_{P(X)}[f(X)] = E_{Q(X)}\left[f(X)\frac{P(X)}{Q(X)}\right]$$

We can easily prove that this equality holds as the following:

$$E_{Q(X)}\left[f(X)\frac{P(X)}{Q(X)}\right] = \sum_{x} Q(x)f(x)\frac{P(X)}{Q(X)}$$
$$= \sum_{x} f(x)P(X)$$
$$= E_P(X)[f(X)]$$

Now, as we have a relationship between the expectations relative to $P(X)$ and $Q(X)$, we can generate samples $D = \{\xi[1], \xi[2], , \xi[M]\}$ from the distribution Q and then estimate the following:

$$\hat{E}_D(f) = \frac{1}{M} \sum_{m=1}^{M} f(x[m])\frac{P(x[m])}{Q(x[m])}$$

We call this the *unnormalized importance sampling estimator*. The main point to note for this estimator is that it's unbiased and hence its mean for any dataset is precisely the desired value. The second thing to note is that the variance of this type of estimator decreases linearly with the number of samples. Hence, we can use this property to estimate the number of samples we need for a good estimate.

One problem with unnormalized importance sampling is that we have considered that we already know P. However, the most frequent reason for sampling from a different distribution Q is that we only know the unnormalized distribution $\tilde{P}(X)$, where $\tilde{P}(X) = ZP(X)$. So in this case, we can define the weightings as follows:

$$w(X) = \frac{\tilde{P}(X)}{Q(X)}$$

With this new weighting, however, our standard estimator for the expectation fails, but we can define a similar estimator for this case as well. Before that, let's see whether the expectation of the random variable $w(X)$ is Z:

$$E_{Q(X)}\left[w(X)\right] = \sum_x Q(X)\frac{\tilde{P}(X)}{Q(X)}$$

$$= \sum_x \tilde{P}(X) = Z$$

Now, we can define our previous estimator function as follows:

$$E_{P(X)}\left[f(X)\right] = \sum_x P(x)f(x)$$

$$= \sum_x Q(x)f(x)\frac{P(x)}{Q(x)}$$

$$= \frac{1}{Z}\sum_x Q(x)f(x)\frac{\tilde{P}(x)}{Q(x)}$$

$$= \frac{1}{Z}E_{Q(X)}\left[f(X)w(X)\right]$$

$$= \frac{E_{Q(X)}\left[f(X)w(X)\right]}{E_{Q(X)}\left[w(X)\right]}$$

Given M samples $D = x[1], x[2],, x[M]$, we can estimate $\tilde{E}_D(f)$ as follows:

$$\tilde{E}_D(f) = \frac{\sum_{m=1}^{M} f(x[m]) w(x[m])}{\sum_{m=1}^{M} w(x[m])}$$

This estimator is known as **normalized importance sampling estimator** or **weighted importance sampling estimator**. Unlike the unnormalized estimator, normalized importance sampling estimators do have a bias.

Importance sampling in Bayesian networks

In this section, we will apply the concept of importance sampling in Bayesian networks. We will discuss the proposal distribution Q, which we usually use in the case of Bayesian networks.

Assume that in a Bayesian network, we want to focus our samples to a particular set of events $Z = z$, either because we want the probability of Z or we have observed Z. Taking the example of our restaurant model, let's say we have observed that the cost is high. It is easy for us to sample the descendant variables of *Cost* according to this condition. However, it is not possible for us to sample the nondescendant variables without performing inference over them.

So now, we define a distribution that simplifies the generation of particles. This new distribution is known as **mutilated network proposal distribution**. Let's say, given a network B and some conditions $Z = z$, we define the mutilated network $B_Z = z$ as follows:

- Each node $Z_i \in Z$ has no parents in , and the CPDs of all $B_Z = z$ give 1 to $Z_i = z_i$ and 0 to all other values $Z_i' \in Val(Z_i)$.
- The parents and CPDs of all other nodes $X \notin Z$ are unchanged.

So, for the case where we observe Cost as high, we get the network as shown in Fig 4.19:

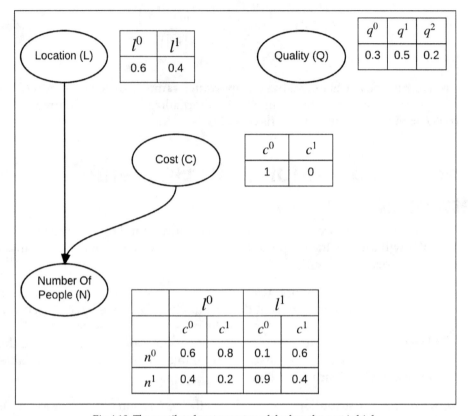

Fig 4.19: The mutilated restaurant model when the cost is high

Importance sampling done with this method is exactly the same as the LW algorithm.

Computing marginal probabilities

If we want to compute the marginal probability of some event $Z = z$, we can simply use forward sampling or do unnormalized sampling with the target distribution as $P_B(X)$ and the proposal distribution Q generated from the mutilated network $B_Z = z$. With the indicator function for our query as $f(\xi) = 1\{\xi(Z) = z\}$, our unnormalized estimator is as follows:

$$\hat{P}_D(z) = \frac{1}{M} \sum_{m=1}^{M} 1\{\xi[m] < Z >= z\} w(\xi[m])$$

$$= \frac{1}{M} \sum_{m=1}^{M} w[m]$$

This equality holds in this case because the samples that have been generated already satisfy $Z = z$.

Ratio likelihood weighting

Now, coming to the problem of computing the conditional probability $P(y \mid e)$, we can use ratio likelihood weighting. We compute $P(y \mid e)$ using the chain rule as $P(y, e) / P(e)$. To compute $P(y, e)$ and $P(e)$, we carry out unnormalized sampling twice, once to generate M particles for $P(y, e)$ and then to generate M' particles for $P(e)$. Then, we use the following equation:

$$\hat{P}_D(y \mid e) = \frac{\hat{P}_D(y, e)}{\hat{P}_{D'}(e)}$$

$$= \frac{\frac{1}{M} \sum_{m=1}^{M} w[m]}{\frac{1}{M'} \sum_{m=1}^{M'} w[m]}$$

Normalized likelihood weighting

Ratio likelihood weighting allowed us to compute the probability of a single query $P(y \mid e)$, but in a case where we want to compute multiple queries or a joint distribution $P(Y \mid e)$, we could use ratio likelihood weighting for each $y \in Y$. This turns out to be computationally very expensive, so we generally use normalized likelihood weighting to compute this.

Markov chain Monte Carlo methods

The LW sampling algorithm correctly samples the posterior of the descendant nodes, but for the nondescendants, it samples the prior and tries to fix it with the weightings. So, for the case where we have most of the observed nodes as leaves of the network, we would be sampling the prior rather than the posterior. We will now discuss an algorithm that generates a sequence of samples. The first samples generated may be near to the prior, but as we keep on generating samples, it keeps getting closer to the posterior. Also, this sampling algorithm works for both Bayesian and Markov networks.

Gibbs sampling

In the Gibbs sampling algorithm, we start by reducing all the factors with the observed variables. After this, we generate a sample for each unobserved variable on the prior using some sampling method, for example, by using a mutilated Bayesian network. After generating the first sample, we iterate over each of the unobserved variables to generate a new value for a variable, given our current sample for all the other variables.

Let's take the example of our restaurant model to make this clearer. Assume that we have already observed that the cost of the restaurant is high. So, we will have the CPDs: $P(L), P(Q), P(c^o \mid L, Q), P(N \mid L, c^o)$. We start by generating our first sample with forward sampling, and let's say our first samples are $l^{(0)} = l^1, q^{(0)} = q^1$ and $n^{(0)} = n^0$. We will now iterate over all of our unobserved variables N, L, Q. Starting with N, we will sample it from the distribution $P(N \mid c^0, l^1)$. As we are computing the distribution over a single variable, we can compute it very easily as follows:

$$
P_\Phi\left(N \mid c^0, l^1\right) = \frac{P\left(l^1\right) P\left(q^1\right) P\left(c^0 \mid l^1, q^1\right) P\left(N \mid l^1, c^0\right)}{\sum_n P\left(l^1\right) P\left(q^1\right) P\left(c^0 \mid l^1, q^1\right) P\left(n \mid l^1, c^0\right)}
$$

$$
= \frac{P\left(N \mid l^1, c^0\right)}{\sum_n \left(n \mid l^1, c^0\right)}
$$

Now that we have sampled $n^{(1)}$ from the distribution $P_\Phi\left(N \mid c^0, l^1\right)$, we continue with the iteration and sample L by conditioning the distribution with the new sample value of N, $n^{(1)}$. Similarly, we go on generating samples.

The thing to notice here is that unlike forward sampling, when sampling here we are taking into consideration the evidences, although this method will not give the true posterior as we began sampling from the prior distribution. Yet, considering the evidence, we are able to generate samples that are much closer to the posterior, and the repetition of this method enables us to keep generating samples that get closer to the posterior distribution.

In the later sections, we will formalize this concept using the Markov chain Monte Carlo method. Using this method, we will be able to generate samples that will be much closer to the posterior distribution.

Markov chains

In the case of graphical models, Markov chains are a graph of states of variables X, and the edges represent the probability of transitioning from one state to another. So, an edge $x \rightarrow x'$ represents the probability of transitioning from the state x to x', represented by $T(x \rightarrow x')$:

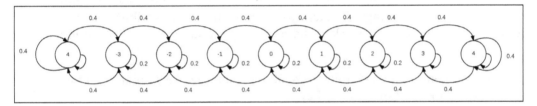

Fig 4.20: Markov chain for a drunk man

Let's take the example of a drunk man walking along a road. The position of the person on the road can be represented by a random variable. Let's say the person started at point 0 and can go ahead to +4 or go behind to -4, but there are walls beyond this point, so even if he tries to go beyond these points he will stay at the same point. Also, the probability of going either forward or backward is 0.4 and the probability of staying in the same position is 0.2, that is, $T(x \rightarrow x+1) = 0.4$, $T(x+1 \rightarrow x) = 0.4$ and $T(x \rightarrow x) = 0.2$ respectively. Also, $T(+4 \rightarrow +4) = T(-4 \rightarrow -4) = 0.4 + 0.2 = 0.6$ as the road is blocked by the walls.

We can consider the position of the man at any given time t to be a random variable represented by $X^{(t)}$. This can be computed as follows:

$$P^{(t+1)}\left(X^{(t+1)} = x'\right) = \sum_{x \in Val(X)} P^{(t)}\left(X^{(t)} = x\right) T\left(x \rightarrow x'\right)$$

Putting the earlier equation in words, we can say that the probability of the person being at point x' at some time $(t + 1)$ is equal to the sum over all the states $x \in Val(X)$ of the product of that person being in that state x and then transitioning to state x' from x.

Let's now try computing a few probability values for the man's position. We know that the man started from the point 0, so at time $t = 0$, $P(X^{(0)} = 0) = 1$. Now, at time $t = 1$, the probability of the man being at point 0 is $P(X^{(1)} = 0) = 0.2$, and the probability of being at +1 or -1 is $P(X^{(1)} = +1) = P(X^{(1)} = -1) = 0.4$. Moving on, at time $t = 2$, the probability of the man being at point 0 is $P(X^{(2)} = 0) = 0.2 * 0.2 + 0.4 * 0.4 + 0.4 * 0.4 = 0.36$, point +1 or -1 is $P(X^{(2)} = +1) = P(X^{(2)} = -1) = 0.4 * 0.2 + 0.2 * 0.4 = 0.16$, and point +2 or -2 is $P(X^{(2)} = +2) = P(X^{(2)} = -2) = 0.4 * 0.4 = 0.16$. We can now see that the probability of being at different states spreads with each time instance, and finally, we will reach a uniform distribution.

To sample from the Markov chain, we can simply select states at each instant of time using the distribution for that instance. However, Markov chains are not a very good method if we want to sample from a uniform distribution, because for the range $[-K, K]$, it takes on average K^2 steps to reach the uniform distribution. So now, let's try to find out when a Markov chain converges and what the distribution on convergence is.

To make the computation simpler, let's take an example of a similar, but much smaller network, as shown in Fig 4.21:

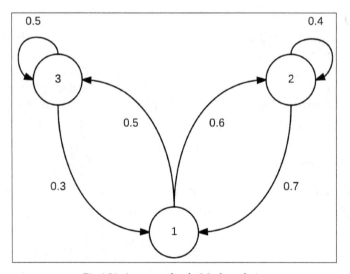

Fig 4.21: An example of a Markov chain

At equilibrium, we can say that for any state x', $p^{(t+1)}$ should almost be equal to $p^{(t)}$:

$$P\left(X^{(t)} = x'\right) = P\left(X^{(t+1)} = x'\right) \approx \sum_{x \in Val(X)} P\left(X^{(t)} = x\right)T\left(x \to x'\right)$$

At equilibrium, the distribution is known as stationary distribution and is represented by $\pi(X)$. We can easily show this as follows:

$$\pi\left(X = x'\right) = \sum_{x \in Val(X)} \pi\left(X = x\right)T\left(x \to x'\right)$$

Now, let's try to compute the stationary distributions for the Markov chain in Fig 4.21. We can write the following equations:

$$\pi\left(x^1\right) = 0.6 * \pi\left(x^2\right) + 0.5 * \pi\left(x^3\right)$$

$$\pi\left(x^2\right) = 0.7 * \pi\left(x^1\right) + 0.4 * \pi\left(x^2\right)$$

$$\pi\left(x^3\right) = 0.3 * \pi\left(x^1\right) + 0.5 * \pi\left(x^3\right)$$

For this to be a legal distribution, it should also satisfy:

$$\pi\left(x^1\right) + \pi\left(x^2\right) + \pi\left(x^3\right) = 1$$

We can now easily solve this set of equations to get the following results:

$$\pi\left(x^1\right) = 0.3615$$

$$\pi\left(x^2\right) = 0.4217$$

$$\pi\left(x^3\right) = 0.2168$$

In this case, we got a unique solution for the distributions, but in general, we cannot guarantee that we will always get a converged distribution. For a finite state Markov chain, we can verify the Markov chain for the following two conditions to check if the distributions converge:

- It is possible to get from any state to another state using a positive probability path
- For each node, there is a single-step positive probability path to get back to it, that is, a self-loop with positive probability

These two conditions are usually sufficient but not necessary to guarantee convergence in the distribution.

The multiple transitioning model

We saw how Markov chains work in cases where we have a single random variable. However, in the case of graphical models, we have multiple variables, and each state of the Markov chain is an assignment to multiple variables. So in this case, it is convenient to decompose our transitioning model so that there is change only in a single variable in each transition. We can extend our drunk man example to understand this better. So now, consider that the man can now go ahead and back as well as left and right. To represent this case with our transitioning model, a pair of random variables will represent the X and Y positions for each state of the Markov chain.

In such cases, we define multiple transitioning models, and each such transitioning model is known as a kernel. Now, to construct the Markov chain from these sets of kernels, we can select a kernel T_i with a probability $\frac{1}{k}$. We could also simply cycle over each of the kernels. However, as we are using different kernels for transitions, our Markov chain can't be homogeneous. To solve this problem, we assume that each transition of the original Markov chain is a combination of k transitions of these kernels.

Using a Markov chain

So far, we have been discussing constructing Markov chains. In this section, we will see how to apply these concepts in the case of our graphical models. In the case of probabilistic models, we usually want to compute the posterior probability $P(Y|E = e)$, and to sample this posterior distribution, we will have to construct a Markov chain whose stationary distribution is $P(Y|E = e)$. So, the states of this Markov chain should be instantiations x of variables $\chi - Y$ and should converge to $\pi(\chi - Y)$.

So, for a state (x_{-i}, x_i) in the Markov chain, we define the kernel T_i as follows:

$$T_i\big((x_{-i}, x_i)(x_{-i}, x_i')\big) = P(x_i \mid x_{-i})$$

We can see that this transition probability doesn't depend on the current value of x_i of X_i but only on the remaining state x_{-i}. Now, it's really easy to show that the posterior distribution $P_\Phi(X) = P(\chi \mid e)$ is a stationary distribution of this process.

In graphical models, Gibbs sampling can be very easily implemented in cases where we can compute the transition probability $P(X_i \mid x_{-i})$ efficiently. We already know the following:

$$P_\Phi = \frac{1}{Z} \prod_j \phi_j(D_j)$$

$$= \frac{1}{Z} \prod_{j : X_i \in D_j} \phi_j(D_j) \prod_{j : X_i \in D_j} \phi_j(D_j)$$

Let $x_{j,-i}$ denote the assignment in x_{-i} to $D_j - \{X_i\}$, noting that when $X_i \notin D_j, x_{j,-i}$ is a full assignment to D_j. We can now derive the following:

$$P(x_i' \mid x_{-i}) = \frac{P(x_i' \mid x_{-i})}{\sum_{x_i''} P(x_i'' \mid x_{-i})}$$

$$= \frac{\dfrac{1}{Z} \prod_{C_j \ni X_i} \phi_j(x_i', x_{j,-i}) \prod_{C_j \not\ni X_i} \phi_j(x_i', x_{j,-i})}{\dfrac{1}{Z} \sum_{x_i''} \prod_{C_j \ni X_i} \phi_j(x_i'', x_{j,-i}) \prod_{C_j \not\ni X_i} \phi_j(x_i'', x_{j,-i})}$$

$$= \frac{\prod_{C_j \ni X_i} \phi_j(x_i', x_{j,-i}) \prod_{C_j \not\ni X_i} \phi_j(x_i', x_{j,-i})}{\sum_{x_i''} \prod_{C_j \ni X_i} \phi_j(x_i'', x_{j,-i}) \prod_{C_j \not\ni X_i} \phi_j(x_i'', x_{j,-i})}$$

$$= \frac{\prod_{C_j \ni X_i} \phi_j(x_i', x_{j,-i})}{\sum_{x_i''} \prod_{C_j \ni X_i} \phi_j(x_i'', x_{j,-i})}$$

Here, we can see that $P\left(x_i', x_{j,-i}\right)$ only requires the factors involving X_i and also vdepends only on the instantiations of the variable x_{-i} of the Markov blanket of X_i. Similarly, in the case of Bayesian networks, this value depends only on the CPDs of X_i and its children.

Collapsed particles

So far, we have discussed various techniques to sample using full instantiations over the variables. However, the problem with full instantiations is that they can only cover a very small region of the space, as the space is exponential to the number of variables. The solution to this is to have partial instantiations of the variables and use a closed-form representation of a distribution over the rest. Collapsed particles are also known as *Rao-Blackwellized* particles.

So, considering $X_p \subset \chi$ as the set of variables over which we will do the assignments and which the particle will depend on, and $X_d \subset \chi$ as the set of variables over which we define a closed-form distribution, if we want to estimate the expectation of some function $f\left(\xi\right)$ relative to our posterior distribution $P\left(X_p, X_d \mid e\right)$ we have the following:

$$E_{P(\xi|e)[f(\xi)]} = \sum_{x_p, x_d} P\left(x_p, x_d \mid e\right) f\left(x_p, x_d \mid e\right)$$

$$= \sum_{x_p} P\left(x_p \mid e\right) \sum_{x_d} P\left(x_d \mid x_p, e\right) f\left(x_p, x_d \mid e\right)$$

$$= \sum_{x_p} P\left(x_p \mid e\right) \left(E_{P\left(X_d \mid x_p, e\right)}\right) \mid f\left(x_p, X_d \mid e\right)$$

Also, we are assuming that the internal expectation can be computed easily. So essentially, we are using a hybrid approach in the case of collapsed particles. We generate particles x_p for the variables X_p and do the exact inference for the variables in X_d. In the case when we have $X_p = \chi$, then we get to the case of full particles. Similarly, when $X_p = \emptyset$, we get to the case of exact inference. Also, as we are doing exact inference on X_d, we are eliminating any bias or variance introduced because of the variables. Therefore, when $\left|X_p\right|$ is small enough, we are able to get much better results using a smaller number of particles.

Collapsed importance sampling

In the case of full particles for importance sampling, we used to generate particles from another distribution, and then, to compensate for the difference, we used to associate a weighting to each particle. Similarly, in the case of collapsed particles, we will be generating particles for the variables X_p and getting the following dataset:

$$D = \left(x_p[m], w[m], P\left(X_d \mid x_p[m], e \right) \right)_{m=1}^{M}$$

Here, the sample $x_p[m]$ is generated from the distribution Q. Now, using this set of particles, we want to find the expectation of $f(\xi)$ relative to the distribution $P(\xi \mid e)$:

$$\hat{E}_D(f) = \frac{\sum_{m=1}^{M} w[m]\left(E_{P\left(X_d \mid x_p[m], e \right)}\left[f\left(x_p[m], X_d, e \right) \right] \right)}{\sum_{m=1}^{M} w[m]}$$

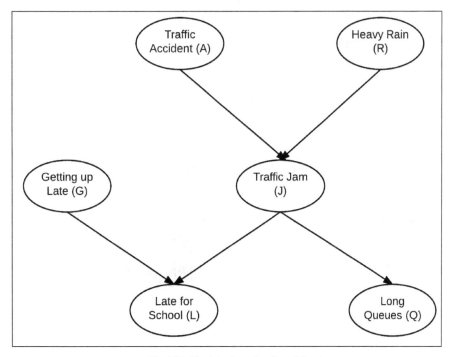

Fig 4.22: The late-for-school model

Let's take an example using the late-for-school model, as shown in Fig 4.22. Let's consider that we have the evidence that j^0, q^1, and partition the variables as $X_P = \{A, R\}$ and $X_D = \{G, L, J, Q\}$. So, we will generate particles over the variable X_P. Also, each such particle is associated with the distribution $P(G, L \mid a, r, j^0, q^1)$. Now, assuming some query (say $P(l^0 \mid j^0, q^1)$), our indicator function will be $P(l^0 \mid j^0, q^1)$. We will now evaluate for each particle:

$$E_{P(G, L \mid a, r, j^0, q^1)} \left[1\{\xi < L >= l^0\} \right] = P(l^0 \mid a, r, j^0, q^1)$$

After this, we will compute the average of these probabilities using the weightings of the samples.

Now, the question is, how do we define the distribution Q and find the weightings for the particles?. We begin by partitioning the evidence variables into two parts, namely E_p and E_d, where $E_p = E \cap X_p$ and $E_d = E \cap X_d$. As the collapsed importance sampling was a hybrid process, we deal with the evidence accordingly, using E_p as evidence in importance sampling and E_d as evidence in exact inference.

Let's consider an arbitrary distribution Q:

$$E_{P(\xi \mid e)} f(\xi) = \sum_{x_p, x_d} P(x_p, x_d \mid e) f(x_p, x_d, e)$$

$$= \sum_{x_p} Q(x_p) \frac{P(x_p \mid e)}{Q(x_p)} \sum_{x_d} P(x_d \mid x_p, e) f(x_p, x_d, e)$$

Using this, we can reformulate $P(x_p \mid e)$ as follows:

$$P(x_p \mid e) = \frac{P(x_p, e)}{P(e)}$$

$$= \frac{P(x_p, e_p, e_d)}{P(e)}$$

$$= \frac{1}{P(e)} P(x_p, e_p) P(e_d \mid x_p, e_p)$$

Let's put this result back into the previous equation:

$$E_{P(\xi|e)}\left[f\left(\xi\right)\right]=\frac{1}{P(e)}\sum_{x_p}Q\left(x_p\right)\frac{P\left(x_p,e_p\right)}{Q\left(x_p\right)}P\left(e_d\mid x_p,e_p\right)\sum_{x_d}P\left(x_d\mid x_p,e\right)f\left(x_p,x_d,e\right)$$

$$=\frac{1}{P(e)}E_{Q(X_p)}\left[\frac{P\left(x_p,e_p\right)}{Q\left(x_p\right)}P\left(e_d\mid x_p,e_p\right)E_{P\left(x_d\mid x_p,e\right)}\left[f\left(x_p,x_d,e\right)\right]\right]$$

From the preceding equation we get the following:

$$w\left(x_p\right)=\frac{P\left(x_p,x_e\right)}{Q\left(x_p\right)}P\left(e_d\mid x_p,e_p\right)$$

Now, computing the mean of importance weights, we get the following estimator:

$$E_{Q(X_P)}\left[w\left(X_P\right)\right]=\sum_{x_p}Q\left(x_p\right)\frac{P\left(x_p\mid e_p\right)}{Q\left(x_p\right)}P\left(e_d\mid x_p,e_p\right)$$

$$=\sum_{x_p}P\left(x_p\mid e_p\right)P\left(e_d\mid x_p,e_p\right)$$

$$=\sum_{x_p}P\left(e_d,x_p,e_p\right)=P\left(e_d,e_p\right)$$

So, we get the final equation as follows:

$$E_{P(\xi|e)}f\left(\xi\right)=\frac{E_{Q(X_P)}\left[w\left(X_P\right)E_{P\left(x_d\mid x_p,e\right)}\left[f\left(x_p,x_d,e\right)\right]\right]}{E_{Q(X_P)}\left[w\left(X_P\right)\right]}$$

In the preceding discussion, we didn't place any restriction on the selection of the distribution Q. The two main points to consider for the selection of the distribution Q are as follows:

- It should be easy to generate samples from this distribution.
- It should be similar to our target distribution $P\left(X_p\mid e\right)$.

In the case of collapsed particles, we will generate particles from the distribution $Q(X_p)$. However, as we saw in the case of full particles, we had to sample over the variable's parents before sampling the variable. In the case of collapsed particles, it is quite possible that the parents of a variable are not in X_p. The simplest solution to this problem is to construct the set X_p in such a way that for every $X \in X_p$, $Par_X \in X_p$ holds as well. To do this, we must use a simple approach to start with the nodes having no parents, include them in , and then work downwards from there.

Summary

In this chapter, we discussed different ways of performing approximate inference in graphical models, such as cluster graph belief propagation, propagation using approximate messages, and inference based on the concepts of sampling from the model. In cluster graph belief propagation, we relaxed the constraint of having a clique tree, and instead, performed belief propagation on the cluster graph. In the propagation using approximate messages, instead of relaxing the constraints on the structure of the graph, we tried to approximate the messages passed between the clusters. Next, we discussed sampling methods in detail. There are two different ways of sampling. The first includes full particles, where each sample has instantiations of all the variables of the network. The other method consists of collapsed particles, where each sample is an instantiation of a subset of the network's variables. We also discussed the problems we face in the case of full particles. In full particles, a very small part of the complete space is covered using each sample, due to which we need many more samples than in the case of collapsed particles. We also discussed the Markov chain Monte Carlo methods that are extensively used in practical problems.

In the next chapter, we will discuss parameter estimation in the case of Bayesian networks. This will help us create graphical models using the data we have.

5
Model Learning – Parameter Estimation in Bayesian Networks

So far in our discussion, we have always considered that we already know the network model as well as the parameters associated with the network. However, constructing these models requires a lot of domain knowledge. In most real-life problems, we usually have some recorded observations of the variables. So, in this chapter, we will learn to create models using the data we have.

To understand this problem, let's say that the domain is governed by some underlying distribution, $P*$. This distribution is induced by the network model, $M* = (K*, \theta*)$. Also, we are provided with a dataset, $D = \{d[1], d[2], ..., d[M]\}$ of M samples. As these data points are obtained from our observations of the actual model, we can say that these data points have been sampled from the distribution, $P*$. Also, we can assume that all the data samples have been independently sampled from the distribution, $P*$. Such data samples are known as **independently and identically distributed (IID)** samples.

Now, we want to select a model from the family of models over the given variables, such that this model, $\tilde{M}$, induces the probability distribution, $P_{\tilde{M}}$, and this distribution is close to the underlying distribution of our domain.

In this chapter, we will discuss the following topics:

- General ideas in learning
- Maximum likelihood parameter estimation
- Bayesian parameter estimation
- Maximum likelihood structure learning
- Bayesian structure learning

General ideas in learning

Before we discuss the specific methods to learn in the graphical models, in this section, we will briefly discuss some general ideas related to learning.

The goals of learning

The perfect solution to our learning task would be to find a model, $\tilde{M}$, so that the probability distribution induced by it is the same as the underlying distribution of our data. However, this is never possible in real life because of computational costs and lack of data. So, as we can't find the exact underlying distribution, we try to optimize our learning task, depending on the goal of learning. To make it clearer, we can think of two different situations. Let's say in the first case, we want to learn the model to answer conditional queries over some specific variables, whereas in the second case, we want to answer multiple queries involving all the variables of the network. Therefore, in the first case, we would like to optimize our learning over variables, over which we want to answer queries at the cost of getting a less-accurate distribution over the other variables. However, in the second case, we want our learned model to be as close to the underlying model as possible, because we have to answer queries over all the variables. Hence, we see that our goal of learning has a huge effect on our learning task.

Density estimation

One of the most common reasons to learn a graphical model is the inference tasks. In this case, we would like our learned model, $\tilde{M}$, to induce a distribution, $P_{\tilde{M}}$, which is as close to the underlying distribution as possible. To measure the distance between these two distributions, we can use the following relative entropy distance measure:

$$D\left(P* \| \ \tilde{P}\right) = E_{\xi \sim P*}\left[log\left(\frac{P*(\xi)}{\tilde{P}(\xi)}\right)\right]$$

However, the problem with this measure is that we also need to know $p*$ to compute the relative entropy. To solve this problem, we simplify the equation as follows:

$$D(P||P') = E_{\xi \sim P}\left[\log\left(\frac{P(\xi)}{P'(\xi)}\right)\right]$$
$$= E_{\xi \sim P}\left[\log P(\xi) - \log P'(\xi)\right]$$
$$= E_{\xi \sim P}\left[\log P(\xi)\right] - E_{\xi \sim P}\left[\log P'(\xi)\right]$$
$$= -H_P(X) - E_{\xi \sim P}\left[\log P'(\xi)\right]$$

Here, we see that the first term depends only on $P*$, and hence, it is unchanged for any choice of model. Therefore, we ignore this term and compare our models only on the basis of the second term, $E_{\xi P*}\left[\log \tilde{P}(\xi)\right]$, and prefer the models that make this term as large as possible. This term is commonly known as **expected log-likelihood**. This term encodes the probability of our model to generate the given data points. Therefore, for a model that has high likelihood value for some given data, it would be closer to our underlying distribution of the data.

So, in our learning problem, we are interested in the likelihood of the data, when a model is given, M, that is, $P(D|M)$. For our convenience, we usually use log-likelihood denoted as $l(D|M) = \log P(D|M)$. We also define log-loss as the negative of log-likelihood. Log-loss is an example of a loss function. A loss function, $loss(\xi|M)$, determines the loss that our model makes on a particular data point, ξ. Therefore, for better learning, we try to find a model that minimizes the expected loss, also known as risk:

$$E_{\xi \sim P*}\left[loss(\xi|M)\right]$$

However, as $P*$ is not known, we can approximate this expected loss by averaging over the sampled data points:

$$E_D\left[loss(\xi|M)\right] = \frac{1}{|D|}\sum_{\xi \in D} loss(\xi|M)$$

Taking the example of log-loss and considering a data set, $D = \{\xi[1], \xi[2], ..., \xi[M]\}$, we have the following equation:

$$P(D|M) = \prod_{m=1}^{M} P(\xi|m|M)$$

Taking the logarithm of the preceding expression, we get the following equation:

$$\log P(D|M) = \prod_{m=1}^{M} \log P(\xi|m|M)$$

As we saw earlier, this term is the negative of the empirical log-loss. Hence, we can easily get a good intuition of empirical risk using log-loss as the loss function.

Predicting the specific probability values

In the preceding section, we tried to learn the complete underlying probability distribution, P^*. For this, we used the log-likelihood function to select the most accurate model. The log-likelihood function uses complete assignments to compute the probability of how likely it is for the data represented by our model. Thus, models learned in this way can be used to answer a whole range of conditional or marginal probability queries over the variables of the model.

In many cases, though, we are more interested in answering a single conditional probability. Let's take the example of a simple classification problem using the Iris dataset for the classification of flower species. We are provided with five variables, namely sepallength, sepalwidth, petallength, petalwidth, and flowerspecies. Now, we want to predict the species of a flower using the sepal length, sepal width, petal length, and petal width of a given flower. So, in this case, we always want to answer a specific conditional distribution over the variables, that is, $P(flowerspecies | sepallength, sepalwidth, petallength, petalwidth)$. Rather in this case, we are interested in the MAP queries over the variable, flowerspecies, when all the other variables are given. In real life, we have a lot of problems like this where we want to answer only some specific queries from our learned model.

Therefore, in such cases, we can select a different loss function that would better represent our problem. For example, in this case, we can use a **classification error**, also known as the **0/1 error**:

$$E_{(x,y)\sim\tilde{P}}\left[1\{h_{\tilde{P}}(x) \neq y\}\right]$$

Here, $1(.)$ is an indicator function; $h_{\tilde{p}}(x)$ represents the predicted value using the hypothesis, $h_{\tilde{p}}$; and y is the actual or target value.

In simple terms, this error function simply computes the probability over all terms sampled from $\tilde{p}$, for which our model selects the wrong label. This error function is good for the case when we want to predict a single variable, or maybe a couple of variables. However, in cases when we want to predict a large number of variables, let's say in the case of image segmentation, we would not like to penalize the whole model for wrongly predicting the value of a single pixel. One suitable error function in such cases is **Hamming loss**, which also does consider the number of variables in each prediction that were predicted wrong.

Therefore, if we know in advance that we are going to use our model for a specific prediction task, we can always optimize our model for those variables.

Knowledge discovery

Another problem that we might want to tackle through learning is that of knowledge discovery, in which we would like to know the relationships between the variables. So, in this case, we mostly focus on predicting the correct network structure. Though, as it turns out, it is very difficult to achieve this with good confidence. So, in the cases where we have a large amount of data, we may be able to construct a network structure with good confidence. In the case of Bayesian networks, there are a lot of I-equivalent structures for any given structure. Therefore, we can at the best hope to learn an I-equivalent structure from the data. Now, coming to the case when we don't have enough data, we will not be able to say anything very confidently about the relationship between the variables. For example, let's say that our data shows a weak correlation between two variables, but as we don't have enough data, we can't confirm this as it might be due to some noise in our data.

Thus, we can conclude that in the case of a knowledge discovery task, the most important thing is to focus on the confidence with which we predict the network structure. In the later sections, we will discuss how to approach such problems.

Learning as an optimization

In the previous sections, we saw various ways of evaluating our models and also defining the loss functions that we want to minimize. This suggests that a learning task can be viewed as an optimizations problem. In an optimization problem, we are provided with a hypothesis space, which in this case, is the set of all possible models along with an objective function, on the basis of which we will select the best-representing model from the hypothesis space. In this section, we will discuss the various choices of objective functions and how they affect our learning task.

Empirical risk and overfitting

Let's consider the task of selecting a model, M, which optimizes the expectation of some loss function, $E_{\xi \sim P*}\left[loss(\xi \mid M)\right]$. As we don't know the value of $P*$, we generally use the dataset, D, which we have to get an empirical estimate of the expectation. Using D, we can define an empirical distribution, $\hat{P}_D$, as follows:

$$\hat{P}_D(A) = \frac{1}{M}\sum_m 1\{\xi[m] \epsilon A\}$$

Putting this in simple words, for some event, A, we assign its probability to be the number of times we have seen this event in our samples. Therefore, as we have more and more samples from the original distribution, $P*$, the value of $\hat{P}_D(A)$ keeps getting closer and closer to the original distribution.

However, there are a few drawbacks to this approach that we need to keep in mind to avoid getting poor results. Think of a case when we have a lot of variables in the network, let's say n. Considering that all the variables can only take two different states, our joint distribution over these variables will have 2^n different assignments. Now, let's say that we are provided with 1000 distinct samples from the original distribution. If we try to find the empirical distribution using this data, we will be assigning a probability of 0.001 to each of the 1000 assignments that were given to us and will assign 0 to the rest $2^n - 1000$ assignments. In real life, we want to predict over new data using our learned model, and it is highly possible that our training data doesn't have all the possible events. In such cases, our trained model will overfit to the training data as it assigns 0 probability to all the events that are not present in the training data.

So, to avoid overfitting, we can limit our hypothesis space to simpler models. This leads to yet another problem; with limited hypothesis space, we might not be able to find a model that will fit perfectly into the original distribution, even if we are provided with infinite data. This type of limitation in learning introduces an inherent error in the learning model, which is known as **bias**. Conversely, if we have a hypothesis space with more complex models, we can correctly learn the actual distribution, $P*$. In that case, if we also have less data, we will get too many fluctuations in our predictions. As a result, we will have a learned model with high variance.

In conclusion, we will always have a trade-off between the bias and variance in our learned models. However, with very limited data, variance turns out to be more dangerous, as it is not able to learn the actual distribution, $P*$, at all.

Discriminative versus generative training

Until now, we have been trying to learn the model to predict all the variables. This kind of learning is known as **generative learning**, as we are trying to generate all the variables, the ones we are trying to predict as well as the ones that we want to use as features. However, as we discussed earlier, in many cases, we already know the conditional distribution that we want to predict. So, in such cases, we try to predict a model so that $\tilde{P}(X|Y)$ is as close as possible to $P*(X|Y)$. This is known as **discriminative learning**.

Learning task

As discussed in the previous sections, we must formalize our learning task. The inputs for our learning task are as follows:

- Constraints for our model, $\tilde{M}$, which will be used to define our hypothesis space

- A set of independent and identically distributed samples, $D = \{d[1], d[2], ..., d[M]\}$, from the original distribution,

The output of our learning will either be the network structure, the parameters, or both. Let's discuss all these in a bit more detail.

Model constraints

Our definition of a hypothesis space depends on several factors. One of the most important factors is the problem that we are trying to solve. There are various kinds of problems that we might face. So, in some cases, we might already know the network structure and might just want to learn the parameters of the network. In some other cases, it is also possible that we want to learn both, the network structure as well as the parameter of the network. Or, it is also possible that we don't even know all the variables of the model that we are trying to learn. In general, the lesser the information we have, the more we have to consider a wider hypothesis space because we need to consider more models to find the one that is the closest.

Other than this, we also discussed how the constraints on the hypothesis space affect the bias and variance in the learned model. One more thing to consider while defining the hypothesis space is the cost of computation, as in a hypothesis space having more complex models, it is generally more difficult to find an optimal model.

Data observability

One other condition that affects our learning task is the extent of observability of our training data. In some cases, it is possible that we might not have the complete data over all the variables, or we might have hidden variables in the actual model, and hence, they may never have been observed.

In many real-life situations, it is not possible to measure all the variables of our model. In such cases, dealing with missing data is critical. We can take the example of designing a model to diagnose a disease in a patient based on some tests. In this case, we can't do all the tests on a patient. Also, we can't say with certainty which disease he has.

Parameter learning

In the previous sections, we have been discussing the general concepts related to learning. Now, in this section, we will be discussing the problem of learning parameters. In this case, we will already know the networks structure and we will have a dataset, $D = \{\xi[1], \xi[2], ..., \xi[M]\}$, of full assignment over the variables. We have two major approaches to estimate the parameters, the maximum likelihood estimation and the Bayesian approach.

Maximum likelihood estimation

Let's take the example of a biased coin. We want to predict the outcome of this coin using previous data that we have about the outcomes of tossing it. So, let's consider that, previously, we tossed the coin 1000 times and we got heads 330 times and got tails 670 times. Based on this observation, we can define a parameter, θ, which represents our chances of getting a heads or a tails in the next toss. In the most simple case, we can have this parameter, θ, to be the probability of getting a heads or tails. Considering θ to be the probability of getting a heads, we have $\theta = 0.33$. Now, using this parameter, we are able to have an estimate of the outcome of our next toss. Also, as we increase the number of data samples that we used to compute the parameter, we will get more confident about the parameter.

Putting this all formally, let's consider that we have a set of independent and identically distributed coin tosses, $D = \{\xi[1], \xi[2], ..., \xi[M]\}$. Each $\xi[i]$ can either take the value heads, *(H)*, with the probability, θ, or tails, *(T)*, with probability, $(1-\theta)$. We want to find a good value for the parameter, θ, so that we can predict the outcomes of the future tosses. As we discussed in the previous sections, we usually approach a learning task by defining a hypothesis space, Θ, and an optimization function. In this case, as we are trying to get the probability of a single random variable, we can define our hypothesis space as follows:

$$\Theta \in [0,1]$$

Now, let's take an example that we have, namely $D = \{T, H, H, T, T, T, H, T\}$. When the value of θ is given, we can compute the probability of observing this data. We can easily say that $P(D[1] = T) = (1-\theta)$. Also, $P(D[2] = H) = \theta$, as all the observations are independent. Now, consider the following equation:

$$P(D|\theta) = (1-\theta)\theta\theta(1-\theta)(1-\theta)(1-\theta)\theta(1-\theta)$$
$$= \theta^3(1-\theta)^5$$

This is the probability of our data to conform with our parameter, θ, which is also known as the likelihood, as we had discussed in the earlier section. Now, as we want our parameter to agree with the data as much as possible, we would like the likelihood, $P(D|\theta)$, to be as high as possible. Plotting the curve of $P(D|\theta)$ within our hypothesis space, we get the following curve:

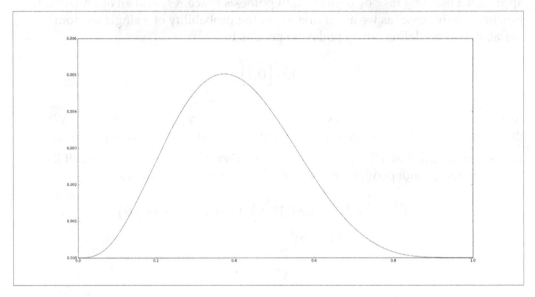

Fig 5.1: Curve showing the variation of likelihood with θ

From the curve in Fig 5.1, we can now easily see that we get the maximum likelihood at $\theta = 0.4$.

Now, let's try to generalize this computation. Also, let's consider that in our dataset, we have M_H number of heads and M_T number of tails:

$$M_H = \sum_{i=0}^{M} 1\{D[i] = H\}$$

$$M_T = \sum_{i=0}^{M} 1\{D[i] = T\}$$

From the example we saw earlier, we can now easily derive the following equation:

$$P(D|\theta) = \theta^{M_H}(1-\theta)^{M_T}$$

Now, we would like to maximize this likelihood to get the most optimum value for θ. However, as it turns out it, it is much easier to work with log-likelihood, and as log-likelihood is monotonically related to the likelihood function, the optimum value of θ for the likelihood function would be the same as that for the log-likelihood function. So, first taking the log of the preceding function, we get the following equation:

$$\log P(D|\theta) = M_H \log \theta + M_H \log(1-\theta)$$

To find the maxima, we now take the derivative of this function and equate it to 0. We get the following result:

$$\frac{M_H}{\hat{\theta}} - \frac{M_T}{\left(1-\hat{\theta}\right)} = 0$$

$$\hat{\theta} = \frac{M_H}{M_H + M_T}$$

Hence, we get our maximum likelihood parameter for the generalized case.

Maximum likelihood principle

In the preceding section, we saw how to apply the maximum likelihood estimator in a simple single variable case. In this section, we will now discuss how to apply this to a broader range of learning problems and how to use this to learn the parameters in the case of a Bayesian network.

Now, let's define our generalized learning problem. We assume that we are provided with a dataset, $D = \{\xi[1], \xi[2], ..., \xi[M]\}$, containing the IID samples over a set of variables, $\mathcal{X}$. We also assume that we know the sample space of the data, that is, we know the variables and the values that it can take. For our learning, we are provided with a parametric model, whose parameters we want to learn. A parametric model is defined as a function, $P(\xi|\theta)$, that assigns a probability to ξ, when a set of parameters is given, θ. As this parametric model is a probability distribution, it should be non-negative and should sum up to 1:

$$\sum_{\xi} P(\xi|\theta) = 1$$

As we have defined our learning problem, we will now move on to applying our maximum likelihood principle on this. So, first of all, we need to define the parameter space for our model. Let's take a few examples to make defining the space clearer.

Let's consider the case of a multinomial distribution, P, which is defined over a set of variables, X, and can take the values, $x^1, x^2, ..., x^K$. The distribution is represented as $\theta \in \mathbf{R}$:

$$P(x|\theta) = \theta_k \qquad if \ x = x^k$$

The parameter space, Θ, for this model can now be defined as follows:

$$\Theta = \theta \in [0,1]^K \ | \sum_i \theta_i = 1$$

We can take another example of a Gaussian distribution on a random variable, X, such that X can take values from the real line. The distribution is defined as follows:

$$P(x|\mu,\sigma) = \frac{1}{\sqrt{2\pi}} e^{-\frac{(x-\mu)^2}{2\sigma^2}}$$

For this model, our parameters are μ and σ. On defining $\theta = <\mu,\sigma>$, our parameter space can be defined as $\Theta = \mathbf{R} \times \mathbf{R}^+$.

Now that we have seen how to define our parameter space, the next step is to define our likelihood function. We can define our likelihood function on our data, D, as $P(D|\theta)$ and it can be expressed as follows:

$$P(D|\theta) = \prod_m P(\xi[m]|\theta)$$

Now, using the earlier parameter space and likelihood functions, we can move forward and compute the maxima of the likelihood or log-likelihood function to find the most optimal value of our parameter, θ. Taking the logarithm of both sides of the likelihood function, we get the following equation:

$$\log P(D|\theta) = \log P(\xi[1]|\theta) + \log P(\xi[2]|\theta) + ... + \log P(\xi[M]|\theta)$$

Now, let's equate this with 0 to find the maxima:

$$\log\Big(P\big(\xi[1]\,|\,\theta\big)\Big) + \log\Big(P\big(\xi[2]\,|\,\theta\big)\Big) + \ldots + \log\Big(P\big(\xi[M]\,|\,\theta\big)\Big) = 0$$

We can then solve this equation to get our desired θ.

The maximum likelihood estimate for Bayesian networks

Let's now move to the problem of estimating the parameters in a Bayesian network. In the case of Bayesian networks, the network structure helps us reduce the parameter estimation problem to a set of unrelated problems, and each of these problems can be solved using techniques discussed in the previous sections.

Let's take a simple example of the network, $X \rightarrow Y$. For this network, we can think of the parameters, θ_{x^0} and θ_{x^1}, which will specify the probability of the variable X; $\theta_{y^1|x^0}$ and $\theta_{y^0|x^0}$, which will specify the probability of $P(Y\,|\,X=0)$, and $\theta_{y^1|x^1}$ and $\theta_{y^0|x^1}$ representing the probability of $P(Y\,|\,X=0)$.

Consider that we have the samples in the form of $\big(x[m], y[m]\big)$, where $x[m]$ denotes assignments to the variable, X, and $y[m]$ denotes assignments to the variable, Y. Using this, we can define our likelihood function as follows:

$$P(D\,|\,\theta) = \prod_{m=1}^{M} P\big(x[m], y[m]\,|\,\theta\big)$$

Utilizing the network structure, we can write the joint distribution, $P(X, Y)$, as follows:

$$P(X,Y) = P(X) \times P(Y\,|\,X)$$

Replacing the joint distribution in the preceding equation using this product form, we get the following equation:

$$P(D|\theta) = \prod_{m=1}^{M} P(x[m]|\theta) P(y[m]|x[m],\theta)$$

$$= \left(\prod_{m=1}^{M} P(x[m]|\theta) \right) \left(\prod_{m=1}^{M} P(y[m]|x[m],\theta) \right)$$

So, we see that the Bayesian network's structure helped us decompose the likelihood function in simpler terms. We now have separate terms for each variable, each representing how well it is predicted, when its parents and parameters are given.

Here, the first term is the same as what we saw in previous sections. The second term can be decomposed further:

$$= \prod_{m=1}^{M} P(y[m]|x[m],\theta_{Y|X})$$

$$= \prod_{m:x[m]=x^0} P(y[m]|x[m],\theta_{Y|X}) \cdot \prod_{m:x[m]=x^1} P(y[m]|x[m],\theta_{Y|X})$$

$$= \prod_{m:x[m]=x^0} P(y[m]|x[m],\theta_{Y|x^0}) \cdot \prod_{m:x[m]=x^1} P(y[m]|x[m],\theta_{Y|x^1})$$

Thus, we see that we can decompose the likelihood function into a term for each group of parameters. Actually, we can simplify this even further. Just consider a single term again:

$$\prod_{m=1} P(y[m]|x[m],\theta_{Y|x^0})$$

These terms can take only two values. When $y[m] = y^0$, it is equal to $\theta_{y^0|x^0}$, and when $y[m] = y^1$, it is equal to $\theta_{y^1|x^0}$. Thus, we get the value, $\theta_{y^0|x^0}$, in cases when $x[m] = x^0$ and $y[m] = y^0$. Let's denote this number by $M\left[x^0, y^0\right]$. Thus, we can rewrite the earlier equation as follows:

$$= \prod_{m:x[m]=x^0} P\left(y[m] | x[m], \theta_{Y|x^0}\right) = \theta\frac{M\left[x^0, y^1\right]}{y^1 | x^0} \cdot \theta\frac{M\left[x^0, y^0\right]}{y^0 | x^0}$$

From our preceding discussion, we know that to maximize the likelihood, we can set the following equation:

$$\theta_{y^1|x^0} = \frac{M\left[x^0, y^1\right]}{M\left[x^0, y^1\right] + M\left[x^0, y^0\right]}$$
$$= \frac{M\left[x^0, y^1\right]}{M\left[x^0\right]}$$

Now, using this equation, we can find all the parameters of the Bayesian network by simply counting the occurrence of different states of variables in the data.

Now, let's see some code examples for how to learn parameters using pgmpy:

```
In [1]: import numpy as np
In [2]: import pandas as pd
In [3]: from pgmpy.models import BayesianModel
In [4]: from pgmpy.estimators import MaximumLikelihoodEstimator

# Generating some random data
In [5]: raw_data = np.random.randint(low=0, high=2, size=(100, 2))
In [6]: print(raw_data)
Out[6]:
array([[1, 1],
       [1, 1],
       [0, 1],
       ...,
       [0, 0],
       [0, 0],
       [0, 0]])
In [7]: data = pd.DataFrame(raw_data, columns=['X', 'Y'])
In [8]: print(data)
Out[8]:
     X  Y
0    1  1
```

```
1    1  1
2    0  1
3    1  0
..   .. ..
996  1  1
997  0  0
998  0  0
999  0  0

[1000 rows x 2 columns]

# Two coin tossing model assuming that they are dependent.
In [9]: coin_model = BayesianModel([('X', 'Y')])
In [10]: coin_model.fit(data,
                        estimator=MaximumLikelihoodEstimator)
In [11]: cpd_x = coin_model.get_cpds('X')
In [12]: print(cpd_x)
Out[12]:
```

x_0	0.46
x_1	0.54

Similarly, we can take the example of the late-for-school model:

```
In [13]: raw_data = np.random.randint(low=0, high=2,

                                     size=(1000, 6))
In [14]: data = pd.DataFrame(raw_data, columns=['A', 'R', 'J',
                                                'G', 'L', 'Q'])

In [15]: student_model = BayesianModel([('A', 'J'), ('R', 'J'),
                                        ('J', 'Q'), ('J', 'L'),

                                        ('G', 'L')])
In [16]: student_model.fit(data,
                          estimator=MaximumLikelihoodEstimator)
In [17]: student_model.get_cpds()
Out[17]:
[<TabularCPD representing P(A: 2) at 0x7f9286b1fa113>,
 <TabularCPD representing P(R: 2) at 0x7f9283b12312>,
 <TabularCPD representing P(G: 2) at 0x7f9383b15114>
 <TabularCPD representing P(J: 2 | A: 2, R: 2) at 0x7f9286bw3329>,
 <TabularCPD representing P(Q: 2 | J: 2) at 0x7f92863kj3294>,

 <TabularCPD representing P(L: 2 | G: 2, J: 2) at

                                     0x7f9282kj49345>]
```

So, learning parameters from data is very easy in `pgmpy` and requires just a call to the `fit` method.

Bayesian parameter estimation

In the preceding section, we discussed the method of estimating parameters using the maximum likelihood, but as it turns out, our maximum likelihood method has a lot of drawbacks. Let's consider the case of tossing a fair coin 10 times. Let's say that we got heads three times. Now, for this dataset, if we go with maximum likelihood, we will have the parameter, $\theta_{head} = 0.3$, but our prior knowledge says that this should not be true. Also, if we get the same results of tossing with a biased coin, we will have the same parameter values. Maximum likelihood fails in accounting for the situation where, because of our prior knowledge, the probability of getting a head in the case of a fair coin should be more or less than in the case of a biased coin, even if we had the same dataset.

Another problem that occurs with a maximum likelihood estimate is that it fails to distinguish between the cases when we get three heads out of 10 tosses and when we get 30000 heads out of 100000 tosses. In both of these cases, the parameter, θ_{heads}, will be 0.3 according to maximum likelihood, but in reality, we should be more confident of this parameter in the second case.

So, to account for these errors, we move on to another approach that uses Bayesian statistics to estimate the parameters. In the Bayesian approach, we first create a probability distribution representing our prior knowledge about how likely are we to believe in the different choices of parameters. After this, we combine the prior knowledge with the dataset and create a joint distribution that captures our prior beliefs, as well as information from the data. Coming back to the example of coin flipping, let's say that we have a prior distribution, $P(\theta)$. Also, from the data, we define our likelihood as follows:

$$P\big(x[m]\,|\,\theta\big) = \begin{cases} \theta & \text{if } x[m]=x^1 \\ 1-\theta & \text{if } x[m]=x^0 \end{cases}$$

Now, we can use this to define a joint distribution over the data, D, and the parameter, θ:

$$P\big(x[1],x[2],...,x[m],\theta\big) = P\big(x[1],x[2],...,x[m]\,|\,\theta\big)P(\theta)$$
$$= P(\theta)\prod_{m=1}^{M}P\big(x[m]\,|\,\theta\big)$$
$$= P(\theta)\theta^{M[1]}(1-\theta)^{M[0]}$$

Here, $M[1]$ is the number of heads in the data and $M[0]$ is the number of tails. Using the preceding equation, we can compute the posterior distribution over θ:

$$P\big(\theta \mid x[1], x[2], ..., x[M]\big) = \frac{P\big(\theta \mid x[1], x[2], ..., x[M] \mid \theta\big) P(\theta)}{P\big(\theta \mid x[1], x[2], ..., x[M]\big)}$$

Here, the first term of the numerator is known as the likelihood, the second is known as the prior, and the denominator is the normalizing factor.

In the case of Bayesian estimation, if we take a uniform prior, it will give the same results as the maximum likelihood approach. So, we won't be selecting any particular value of θ in this case. We will try to predict the outcome of the next coin toss, when all the previous data samples are given:

$$P\big(x[M+1]x[1], x[2], ..., x[M]\big) =$$

$$\int P\big(x[M+1] \mid \theta, x[1], x[2], ..., x[M]\big) P\big(\theta \mid x[1], x[2], ..., x[M]\big) d\theta$$

$$= \int P\big(x[M+1] \mid \theta\big) P\big(\theta \mid x[1], x[2], ..., x[M]\big) d\theta$$

In simple words, here we are integrating our posterior distribution over θ to find the probability of heads for the next toss.

Now, applying this concept of the Bayesian estimator to our coin tossing example, let's assume that we have a uniform prior over θ, which can take values in the interval, $[0, 1]$. Then, $= P\big(\theta \mid x[1], x[2], ..., x[M]\big)$ will be proportional to the likelihood, $P\big(x[1], x[2], ..., x[M] \mid \theta\big) = \theta^{M[1]} (1-\theta)^{M[0]}$. Let's put this value in the integral:

$$P\big(X(M+1) = x^1 \mid x[1], x[2], ..., x[M]\big) = \frac{1}{P\big(x[1], x[2], ..., x[M]\big)} \int \theta \cdot \theta^{M[1]} (1-\theta)^{M[0]} \cdot d\theta$$

Solving this equation, we finally get the following equation:

$$P\big(X(M+1)=x^{1}\mid x[1],x[2],...,x[M]\big)=\frac{M[1]+1}{M[1]+M[0]+2}$$

This prediction is known as the Bayesian estimator. We can clearly see from the preceding equation that as the number of samples increase, the parameters comes closer and closer to the maximum likelihood estimate. The estimator that corresponds to a uniform prior is often referred to as Laplace's correction.

Priors

In the preceding section, we discussed the case when we have uniform priors. As we saw, in the case of uniform priors, the estimator is not very different from the maximum likelihood estimator. Therefore, in this section, we will move on to discuss the case when we have a non-uniform prior. We will show an example over our coin tossing example, using our prior to be a Beta distribution.

A Beta distribution is defined in the following way:

$$\theta \sim Beta(\alpha_{1},\alpha_{0}) \quad if \quad p(\theta)=\gamma\theta^{\alpha_{1}-1}(1-\theta)^{\alpha_{0}-1}$$

Here, α_{0} and α_{1} are the parameters, and the constant, γ, is a normalizing constant, which is defined as follows:

$$\gamma=\frac{\Gamma(\alpha_{1}+\alpha_{0})}{\Gamma(\alpha_{1})\Gamma(\alpha_{0})}$$

Here, the gamma function, $\Gamma(x)$, is defined as follows:

$$\Gamma(x)=\int_{0}^{\infty}t^{x-1}e^{-t}.dt$$

For now, before we start our observations, let's consider that the hyper parameters, α_0 and α_1, correspond to the imaginary number of tails and heads. To make our statement more concrete, let's consider the example of a single coin toss and assume that our distribution, $P(\theta) = Beta(\alpha_0, \alpha_1)$. Now, let's try to compute the marginal probability:

$$P(X[1] = x^1) = \int_0^1 P(X[1] = x^1 \mid \theta).P(\theta)\,d\theta$$

$$= \int_0^1 \theta \cdot P(\theta)\,d\theta$$

$$= \frac{\alpha_1}{\alpha_1 + \alpha_0}$$

So, this conclusion shows that our statement about the hyper parameters is correct. Now, extending this computation for the case when we saw $M[1]$ heads and $M[0]$ tails, we get the following equation:

$$P(\theta \mid x[1], x[2], \ldots, x[M]) \propto P(x[1], x[2], \ldots, x[M] \mid \theta) P(\theta)$$

$$\propto \theta^{M[1]}(1-\theta)^{M[0]} \cdot \theta^{\alpha_1 1}(1-\theta)^{\alpha_0 - 1}$$

$$= \theta^{\alpha_1 + M[1] - 1}(1-\theta)^{\alpha_0 + M[0] - 1}$$

$$= Beta\left(\alpha_1 + M[1] - 1(1-\theta)^{\alpha_0 + M[0] - 1}\right)$$

This equation shows that if the prior distribution is a Beta distribution, the posterior distribution also turns out to be a Beta distribution. Now, using these properties, we can easily compute the probability over the next toss:

$$P(X[M+1] = x^1 \mid x[1], x[2], \ldots, x[M]) = \frac{\alpha_1 + M[1]}{\alpha + M}$$

Here, $\alpha = \alpha_1 + \alpha_0$, and this posterior represents that we have already seen $\alpha_1 + M[1]$ heads and $\alpha_0 + M[0]$ tails.

Bayesian parameter estimation for Bayesian networks

Again, let's take our simple example of the network, $X \rightarrow Y$, and our training data, $D = \{< X[1], Y[1] >, < X[2], Y[2] >, \ldots, < X[M], Y[M] >\}$. We also have unknown parameters, θ_X and $\theta_{Y|X}$. We can think of a dependency network over the parameters and data, as shown in Fig 5.2.

This dependency structure gives us a lot of information about datasets and our parameters. We can easily see from the network that different data instances are independent of each other if the parameters are given. So, $X[m]$ and $Y[m]$ are d-separated from $X[m']$ and $Y[m']$ when θ_X and $\theta_{X|Y}$ are given.

Also, when all the $x[m]$ and $y[m]$ values are observed, the parameters, θ_X and $\theta_{X|Y}$, are d-separated. We can very easily prove this statement, as any path between θ_X and $\theta_{X|Y}$ is in the following form:

$$\theta_X \rightarrow X[m] \rightarrow Y[m] \leftarrow \theta_{Y|X}$$

When $X[m]$ and $Y[m]$ are observed, influence cannot flow between θ_X and $\theta_{Y|X}$. So, if these two parameters are independent a priori then they will also be independent a posteriori. This d-separation condition leads us to the following result:

$$P(\theta_X, \theta_{Y|X} \mid D) = P(\theta_X \mid D) P(\theta_{Y|X} \mid D)$$

This condition is similar to what we saw in the case of the maximum likelihood estimation. This will allow us to break up the estimation problem into smaller and simpler problems, as shown in the following figure:

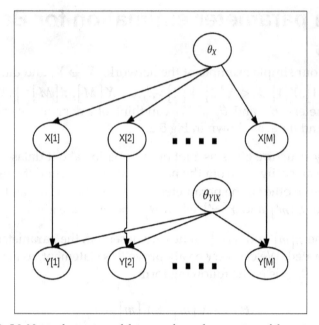

Fig 5.2: Network structure of data samples and parameters of the network

Now, using the preceding results, let's formalize our problem and see how the results help us solve it. So, we are provided with a network structure, G, whose parameters are θ. We need to assign a prior distribution over the network parameters, $P(\theta)$. We define the posterior distribution over these parameters as follows:

$$P(\theta \mid D) = \frac{P(D \mid \theta)P(\theta)}{P(D)}$$

Here, the term, $P(\theta)$, is our prior distribution, $P(D \mid \theta)$ is the likelihood function, $P(\theta \mid D)$ is our posterior distribution, and $P(D)$ is the normalizing constant.

As we had discussed earlier, we can split our likelihood function as follows:

$$P(D \mid \theta) = \prod_i P\left(D \mid \theta_{X_i \mid Pa_{X_i}}\right)$$

Also, let's consider that our parameters are independent:

$$P(\theta) = \prod_i P\left(\theta_{X_i \mid Pa_{X_i}}\right)$$

Combining these two equations, we get the following equation:

$$P(\theta \mid D) = \frac{1}{P(D)} \prod_i \left[P_i\left(D \mid \theta_{X_i \mid Pa_{X_i}}\right) P\left(\theta_{X_i \mid Pa_{X_i}}\right) \right]$$

In the preceding equation, we can see that each of the product terms is for a local parameter value. With this result, let's now try to find the probability of a new data instance given our previous observations:

$$P\left(x[M+1], y[M+1] \mid D\right) = \int P\left(x[M+1], y[M+1] \mid D, \theta\right) P(\theta \mid D) d\theta$$

As we saw earlier, all the data instances are independent. If the parameter is given, we get the following equation:

$$P\left(x[M+1], y[M+1] \mid D, \theta\right)$$
$$P\left(x[M+1], y[M+1] \mid \theta\right)$$
$$P\left(x[M+1] \mid \theta_X\right) P\left(y[M+1] \mid x[M+1], \theta_{Y\mid X}\right)$$

We can also decompose the posterior probability as follows:

$$P\left(x[M+1], y[M+1] \mid D\right)$$

$$= \iint P\left(x[M+1] \mid \theta_X\right) P\left(y[M+1] \mid x[M+1], \theta_{Y\mid X}\right) P(\theta_X \mid D) P(\theta_Y \mid D) d\theta_X d\theta_Y$$

$$= \left(\int P\left(x[M+1] \mid \theta_X\right) P(\theta_X \mid D) d\theta_X \right) \left(\int P\left(y[M+1] \mid x[M+1], \theta_{Y\mid X}\right) P(\theta_Y \mid D) d\theta_Y \right)$$

Now, using this equation, we can solve the prediction problem for each of the variables separately.

Now, let's see some examples of the network's learning parameters using this Bayesian approach on the late-for-school model:

```
In [1]: import numpy as np
In [2]: import pandas as pd
In [3]: from pgmpy.models import BayesianModel
In [4]: from pgmpy.estimators import BayesianEstimator

# Generating some random data
In [5]: raw_data = np.random.randint(low=0, high=2,

                                     size=(1000, 6))
In [6]: print(raw_data)
Out[6]:
array([[1, 0, 1, 1, 1, 0],
       [1, 0, 1, 1, 1, 1],
       [0, 1, 0, 0, 1, 1],
       ...,
       [1, 1, 1, 0, 1, 0],
       [0, 0, 1, 1, 0, 1],
       [1, 1, 0, 0, 1, 1]])
In [7]: data = pd.DataFrame(raw_data, columns=['A', 'R', 'J',
                                               'G', 'L', 'Q'])

# Creating the network structures
In [8]: student_model = BayesianModel([('A', 'J'), ('R', 'J'),
                                       ('J', 'Q'), ('J', 'L'),

                                       ('G', 'L')])
In [9]: student_model.fit(data, estimator=BayesianEstimator)
In [10]: student_model.get_cpds()
Out[10]:
[<TabularCPD representing P(A: 2) at 0x7f92892304fa>,
 <TabularCPD representing P(R: 2) at 0x7f9286c9323b>,
 <TabularCPD representing P(G: 2) at 0x7f9436c9833b>,
 <TabularCPD representing P(J: 2 | A: 2, R: 2) at 0x7f9286s23a34>,
 <TabularCPD representing P(L: 2 | J: 2, G: 2) at

0x7f9286a932b30>,
 <TabularCPD representing P(Q: 2 | J: 2) at 0x7f9286d12904>]

In [11]: print(student_model.get_cpds('D'))
Out[11]:
```

```
| D_0 | 0.44|
| D_1 | 0.56|
```

Therefore, to learn the data using the Bayesian approach, we just need to pass the estimator type `BayesianEstimator`.

Structure learning in Bayesian networks

In the previous sections, we considered that we already know the network structure and we tried to estimate the parameters of the network using the data. However, it is quite possible that we might neither know the network structure nor have the domain knowledge to construct the network. Hence, in this section, we will discuss constructing the model structure when the data is given.

Constructing the model from the data is a difficult problem. Let's take an example of tossing two coins and representing the outcome of the first with the variable, X, and the second with the variable, Y. We know that if the coins are fair, these two random variables should be independent of each other. However, to get this independence condition just from the data, we need to have all these outcomes equal number of times in the data that we will rarely see in real life.

So, in general, we need to make some assumptions about the dependencies. The assumptions that we make will largely depend on our learning task, as we discussed in the previous sections. Now, in the case of knowledge discovery, we would like to know the dependencies between the variables; therefore, we would like our network structure to be as accurate as possible. We know that each distribution can have many P-maps; therefore, the best we can do is get an I-equivalent structure of the original network, $G*$. As we mentioned earlier, it is really hard to get the exact network structure, so we will often have a situation where we have to decide whether we want to include a less-probable edge in our model or not. Although, if we include too many or very few edges, we will end up not learning a good model. Therefore, this decision usually depends on our application.

Other than knowledge discovery, we very often use the models for density estimation. In this case, we would like to learn a model that should be able to learn the underlying probability distribution, $P*$, using our model, and should be able to make predictions over new data points. We may think that in this case, adding less-probable edges to the model will help us learn, as we should be able to learn $P*$ using a more complex model. However, it turns out that our intuition is wrong in this case.

Let's get back to our two-coin tossing example and consider that our dataset consists of 50 samples with the following frequencies: 11 heads/heads, 10 heads/tails, 14 tails/heads, and 15 tails/tails. As all the frequencies are not equal, the data suggests that the two variables, X and Y, are not independent. So, let's consider that while learning using this data, we added an edge between X and Y. In this case, using the maximum likelihood estimator, we get the following parameters: $P(X = H) = 0.42$, $P(X = T) = 0.58$, $P(Y = H | X = H) = 0.22$, $P(Y = T | X = H) = 0.2$, $P(Y = H | X = T) = 0.28$, and $P(Y = T | X = T) = 0.3$. Whereas, in a case when we do not consider any edges between X and Y, we get the following parameters for Y: $P(Y = H) = 0.5$ and $P(Y = T) = 0.5$. It was definitely possible for the probabilities to be skewed, even when we didn't consider any edges between the two variables. In the case of a more complex model, it is more probable that the parameters will be more skewed. This happens because in more complex models, we have lesser data to compute the parameters because of the conditioning. So, for example, while computing $P(Y = T | X = H)$, we will only consider data samples for which $X = H$. Hence, we are left only with 29 samples; whereas in the case of the model when we had no edges, we considered all 50 samples for computing the probability values.

Hence, it's often better to consider simpler models in the case of density estimation problems. A simpler model might not be able to represent the underlying distribution, P^*, very well, but it can be a much better model to generalize over the dataset and will give much better results on new data points.

Methods for the learning structure

In general, there are are three main ways to learn structure. We will be giving a short introduction to each of them in this section; we will go into details in the later sections.

- **Constraint-based structure learning**: The constraint-based structure learning method works on the basis of considering a Bayesian network to be a set of dependence conditions between the random variables. So, in this method, we try to find the dependence conditions from the data given to us. Using these conditions, we then try to construct a network. One major drawback of this method is that if we get a wrong result from our dependence tests, our whole learning fails.

- **Score-based structure learning**: In this method, we consider the Bayesian network as a statistical model. We then define a hypothesis space of possible structures and a scoring function that tells us how close our structure is to the underlying structure. Based on these results, we then try to select the model that represents our underlying structure most closely. As this learning method considers the whole model at once, it is able to give better results than constraint-based learning. The problem with this model is that as our hypothesis space can be very large, finding the most optimal structure is hard. Hence, we generally resort to heuristic search techniques.

- **Bayesian model averaging**: In this method, we try to apply concepts similar to the ones we saw in earlier sections to learn many structures, and then use an ensemble of all these structures. As the number of network structures can be huge, we sometimes have to use some approximate methods to do this.

Constraint-based structure learning

In this method, we try to construct the network structure using the independence conditions obtained from the data. In other words, we try to construct a minimal I-map, given the independence conditions.

Hence, once we have the independence conditions, we can construct a network structure using the algorithm that we discussed earlier, but how do we answer these independence queries from our data?

As we can expect, this question has been extensively studied in statistics and there are numerous methods to answer such queries. We will discuss one of these queries, which is known as the hypothesis testing method. We know that if two random variables are independent, they should satisfy the following condition:

$$P(X, Y) = P(X) \cdot P(Y)$$

So, in our case, we would like to check whether $P^*(X, Y) = P^*(X) \cdot P^*(Y)$. However, in real-life problems, we don't know $P^*(X)$ and $P^*(Y)$, and therefore, we use the following equation to check our hypothesis:

$$P^*(X, Y) = \hat{P}(X) \cdot \hat{P}(Y)$$

Now, using the data samples, we can check whether this equation holds for our data or not. To do this, we will need a decision rule that will tell us whether the two variables are independent given the data samples.

A decision rule should basically compare the two distributions and be able to give a result on whether the independence holds or not. If we go for a very liberal decision rule, it will return the two variables to be independent even when they are not. Similarly, if we consider a very tight-bound decision rule, it will result in saying that the two variables are dependent even when they are independent. A standard way to design a decision function is to measure the distribution's deviance from the independence condition.

For two random variables, X and Y, to be independent, we can expect their count, $M[x, y]$ in the dataset to be somewhere around $M \cdot P(x) \cdot P(y)$. Here, M is the total number of data samples that we have. Specially, in the cases when M is large, this condition should be satisfied. Based on this intuition, we will now derive a deviance measure, commonly known as χ^2 statistic:

$$d_{\chi^2}(D) = \sum_{x,y} \frac{\left(M[x, y] - M \cdot \hat{P}(x) \cdot \hat{P}(y)\right)^2}{M \cdot \hat{P}(x) \cdot \hat{P}(y)}$$

We can clearly see here that when our data fits our independence assumptions perfectly, $d_{\chi^2}(D) = 0$, and the farther it is from our assumption, the greater value it returns.

Another deviance measure technique based on counts is mutual information, $I_{\hat{P}_D}(X; Y)$, and is defined as follows:

$$d_I(D) = I_{\hat{P}_D}(X; Y) = \frac{1}{M} \sum_{x,y} M[x, y] \log \frac{M[x, y]}{M[x]M[y]}$$

Now, using any such deviance measure, we can define a threshold based on which our decision function will make decisions on whether two random variables are independent or not:

$$R_{d,t}(D) = \begin{cases} \textit{independent} & \textit{if } d(D) <= t \\ \textit{dependent} & \textit{if } d(D) > t \end{cases}$$

So, simply put, if the deviance measure is more than the threshold that we have given, the decision function will return the variables as dependent, and if not, they will be independent.

Structure score learning

As we discussed earlier, the score-based method uses a score function to score all the structures in our hypothesis space. Then, using the scores of all the models, we try to select the most optimal structure. So, in this learning method, the most important decision that we have to make is which scoring function to choose. Let's discuss two of the most commonly-used scoring functions.

The likelihood score

As we discussed earlier, the likelihood function gives us the probability of our data given the parameters of the model. So, we would like to select a model that has the maximum likelihood. In the case of structure learning, we want to learn both, the structure and the parameters of the structure. Therefore, our hypothesis space would be much larger than what we saw in the case of parameter learning.

Let's denote our graph and its parameters as $< G, \theta_G >$. Now, we want to find the value using the following equation:

$$\max_{G, \theta_G} P\left(D \mid < G, \theta_G >\right) = \max_G \left[\max_{\theta_G} P\left(D \mid < G, \theta_G >\right) \right]$$

$$= \max_G \left[P\left(D \mid < G, \hat{\theta}_G >\right) \right]$$

Here, $\hat{\theta}_G$ represents the maximum likelihood parameters for the graph, G. Therefore, in simple words, we want to find a graph that has the maximum likelihood when we use the maximum likelihood parameters for it.

To get more insight on this method, let's take the previous example of tossing two coins. So, there are two possibilities for the network structure. One where both the random variables, X and Y, are independent, and thus have no edges between them. The other possibility is to have a network structure, where X is the parent to Y, that is, $X \rightarrow Y$. Considering the network of the independent case to be G_0, we can get its likelihood score as follows:

$$P\left(D \mid G_0\right) = \sum_m \log \hat{\theta}_{x[m]} + \log \hat{\theta}_{y[m]}$$

Considering the other model, $X \rightarrow Y$, as G_1, we can write its likelihood score as follows:

$$P(D|G_1) = \sum_m \log \hat{\theta}_{x[m]} + \log \hat{\theta}_{y[m]x[m]}$$

Here, the $\hat{\theta}$ values are the maximum likelihood estimates. Let's consider the difference of these likelihood scores:

$$P(D|G_1) - P(D|G_0) = \sum_m \log \hat{\theta}_{y[m]x[m]} - \log \hat{\theta}_{y[m]}$$

$$P(D|G_1) - P(D|G_0) = \sum_{x,y} M[x,y] \log \hat{\theta}_{y|x} - \sum_y M[y] \log \hat{\theta}_y$$

Now, let $\hat{P}$ be the empirical distribution over the data. Therefore, we can say that $M[x,y] = M \cdot \hat{P}(x,y)$ and also $M[y] = M \cdot \hat{P}(y)$. Also, we have $\hat{\theta}_{y|x} = \hat{P}(y|x)$ and $\hat{\theta}_y = \hat{P}(y)$. Replacing these values in the preceding equation, we get the following equation:

$$P(D|G_1) - P(D|G_0) = M \sum_{x,y} \hat{P}(x,y) \log \frac{\hat{P}(y|x)}{\hat{P}(y)}$$

$$= M \cdot I_{\hat{P}}(X;Y)$$

Here, $I_{\hat{P}}(X;Y)$ is the mutual information between X and Y in the distribution, $\hat{P}$. Hence, we see here that a higher mutual information means there is a stronger connection between the variables, X and Y, and therefore, the model, G_1, is the more optimal choice.

We can actually generalize this for general networks. We already know that we can write the log-likelihood function as follows:

$$\log P\left(D \mid \hat{\theta}_G\right) = \sum_{i=1} n \left[\sum_{u_i \in Val\left(Pa_{X_i}^G\right)} \sum_{x_i} M\left[x_i, u_i\right] \log \hat{\theta}_{x_i \mid u_i} \right]$$

Let's consider a single term from the earlier equation and $U_i = Pa_{X_i}$:

$$\frac{1}{M} \sum_{u_i} \sum_{x_i} M\left[x_i, u_i\right] \log \hat{\theta}_{x_i \mid u_i} = \sum_{u_i} \sum_{x_i} \hat{P}\left(x_i, u_i\right) \log \hat{P}\left(x_i \mid u_i\right)$$

$$= \sum_{u_i} \sum_{x_i} \hat{P}\left(x_i, u_i\right) \log \left(\frac{\hat{P}\left(x_i, u_i\right)}{\hat{P}\left(u_i\right)} \frac{\hat{P}\left(x_i\right)}{\hat{P}\left(x_i\right)} \right)$$

$$= \sum_{u_i} \sum_{x_i} \hat{P}\left(x_i, u_i\right) \log \frac{\hat{P}\left(x_i, u_i\right)}{\hat{P}\left(x_i\right) \hat{P}\left(u_i\right)} + \sum_{x_i} \left(\sum_{u_i} \hat{P}\left(x_i, u_i\right) \right) \log \hat{P}\left(x_i\right)$$

$$= I_{\hat{P}}\left(X_i; U_i\right) - \sum_{x_i} \hat{P}\left(x_i\right) \log \frac{1}{\hat{P}\left(x_i\right)}$$

$$= I_{\hat{P}}\left(X_i; U_i\right) - H_{\hat{P}}\left(X_i\right)$$

Here, the mutual information is, $I_{\hat{P}}\left(X_i; U_i\right) = 0$ when $Pa_{X_i} = \emptyset$. Also, the second term in the equation, $H_{\hat{P}}\left(X_i\right)$, doesn't depend on the network structure, and therefore, we can ignore this term when comparing the likelihoods of models.

This result tells us that the likelihood score of the structure measures the strength of the dependencies of the variables and their parents.

In this section, until now, we have seen how the likelihood score works. In the case of generalizing the model for newer data points, the likelihood score gives poor results. We can take the example of tossing two coins. As we saw earlier, the difference of their likelihoods is as follows:

$$P\left(D \mid G_1\right) - P\left(D \mid G_0\right) = M \cdot I_{\hat{P}}\left(X; Y\right)$$

As we know, the mutual information between two variables is always non-negative. Hence, the likelihood score of the network, $X \rightarrow Y$, will always be higher than the case when the two variables are independent. Hence, we see that the likelihood score always gives preference to more complex models over simpler models.

Also, as we never have completely independent variables in our data samples because of the added noise, likelihood scores will always select a fully-connected graph over all the variables, as it would be the most complex structure and, hence, would overfit the training data and not give good prediction results over new queries.

Let's see an example of tossing two coins using pgmpy:

```
In [1]: import numpy as np
In [2]: import pandas as pd
In [3]: from pgmpy.models import BayesianModel
In [4]: from pgmpy.estimators import MaximumLikelihoodEstimator

# Generating random data
In [5]: raw_data = np.random.randint(low=0, high=2,
                                      size=(1000, 2))
In [6]: data = pd.DataFrame(raw_data, columns=['X', 'Y'])

In [7]: coin_model = BayesianModel()
In [8]: coin_model.fit(data, estimator=MaximumLikelihoodEstimator)

In [9]: coin_model.get_cpds()
Out[9]:
[<TabularCPD representing P(X: 2) at 0x7f57bd99a588>,
 <TabularCPD representing P(Y: 2 | X: 2) at 0x7f57bd99a198>]

In [10]:coin_model.get_nodes()
Out[10]: ['X', 'Y']

In [11]: coin_model.get_edges()
Out[11]:[('X', 'Y')]
```

The Bayesian score

In the preceding section, we saw scoring based on likelihood and also saw how it is prone to overfitting. Now, in this section, we will discuss another method of scoring from a Bayesian perspective. As we saw in the case of parameter learning, we will have to assign prior probabilities in this case as well. So, we will assign a prior probability, *P(G)*, to the structure of the network, and a prior probability, $P(\theta|G)$, to the parameters of this network structure.

From the Bayes' rule, we know the following equation:

$$P(G \mid D) = \frac{P(D \mid G) \cdot P(G)}{P(D)}$$

Here again, the denominator is just a normalizing factor, and therefore, we will ignore it and define the Bayesian score as follows:

$$\log P(G \mid D) = \log P(D \mid G) + \log P(G)$$

The addition of a prior distribution term in the scoring function allows us to have control over the complexity of the model. Therefore, we assign smaller prior values on more complex models, and thus, we are able to penalize the complex models.

The other term in our scoring function, $\log P(D \mid G)$, takes care of the uncertainty in the parameters:

$$P(D \mid G) = \int_{\Theta_G} P(D \mid \theta_G, G) P(\theta_G \mid G) d\theta_G$$

Here, $P(D \mid \theta_G, G)$ is the likelihood of the data, when a network and its parameters is given, and $P(\theta_G \mid G)$ is our prior distribution over different values of θ for a given network, G.

The Bayesian approach does tell us that the parameter, $\hat{\theta}$, is most probable when the dataset D is given. However, the posterior also gives us a range of choices on how likely each of these is. By integrating $P(D \mid \theta_G, G)$ over θ_G, we are thus measuring the expected likelihood over our parameters, θ_G.

Now, let's see how to compute the marginal likelihoods in simpler cases. Let's consider a single random variable, X, with a prior distribution, $Dirichlet(\alpha_1, \alpha_0)$. Also, consider that our data set contains M[1] heads and M[0] tails. The maximum likelihood is as follows:

$$P(D \mid \hat{\theta}) = \left(\frac{M[1]}{M} \right)^{M[1]} \cdot \left(\frac{M[0]}{M} \right)^{M[0]}$$

Now, let's consider the marginal likelihood. We need to compute the probability over our data, $P(X[1], X[2],..., X[M])$, given the prior. Using the chain rule, we have the following equation:

$$P(x[1], x[2],..., x[M]) = P(x[1]) \cdot P(x[2] | x[1]).....P(x[M] | x[1], x[2],..., x[M-1])$$

Using the Beta prior, we have the following equation:

$$P(x[m+1] | x[1], x[2],..., x[m]) = \frac{M[1]^m + \alpha_1}{m + \alpha}$$

Here, $M[1]^m$ is the number of heads in the first m samples of the dataset. We can take an example of the dataset, $D = \langle H, T, T, H, H \rangle$:

$$P(x[1], x[2],..., x[5]) = \frac{\alpha_1}{\alpha} \cdot \frac{\alpha_0}{\alpha+1} \cdot \frac{\alpha_0 + 1}{\alpha+2} \cdot \frac{\alpha_1 + 1}{\alpha+3} \cdot \frac{\alpha_1 + 2}{\alpha+4}$$

Using the values, $\alpha_1 = \alpha_0 = 1$ and $\alpha = \alpha_1 + \alpha_0 = 2$, we have the following equation:

$$\frac{[1 \cdot 2 \cdot 3] \cdot [1 \cdot 2]}{2 \cdot 3 \cdot 4 \cdot 5 \cdot 6} = \frac{12}{720} = 0.017$$

This value is significantly lower than the likelihood.

In general, for a binomial distribution with a Beta prior, we have the following equation:

$$P(x[1], x[2],..., x[M]) = \frac{\left[\alpha_1...(\alpha_1 + M[1]-1)\right]\left[\alpha_0 ...(\alpha_0 + M[0]-1)\right]}{\alpha...(\alpha + M - 1)}$$

Note here that all the terms inside the square brackets are the products of a sequence of numbers. If α is an integer, we can write this term as $\frac{(\alpha + M - 1)!}{(\alpha_1)!}$, but in this case, we don't know whether α is an integer. So, we use a generalized gamma function to represent such terms:

$$\alpha(\alpha+1)...(\alpha + M - 1) = \frac{\Gamma(\alpha + M)}{\Gamma(\alpha)}$$

Using this result in our earlier equation, we get the following equation:

$$P(x[1], x[2],..., x[M]) = \frac{\Gamma(\alpha)}{\Gamma(\alpha+M)} \cdot \frac{\Gamma(\alpha_1 + M[1])}{\Gamma(\alpha_1)} \cdot \frac{\Gamma(\alpha_0 + M[0])}{\Gamma(\alpha_0)}$$

We can have a generalized formula for multinomial distributions as well:

$$P(x[1], x[2],..., x[M]) = \frac{\Gamma(\alpha)}{\Gamma(\alpha+M)} \cdot \prod_{i=1}^{k} \frac{\Gamma(\alpha_i + M[x^i])}{\Gamma(\alpha_i)}$$

The Bayesian score for Bayesian networks

In the preceding section, we discussed computing the Bayesian score in the case of single random variables. In this section, we will generalize our discussion to compute the Bayesian score for Bayesian networks. Again, we will take the case of having two random variables, X and Y, and two possible network structures over them. We will denote the structure with no edges between X and Y with G_0 and the $X \rightarrow Y$ network with G_1.

For G_0, we have the following equation:

$$P(D|G_0) = \int_{\Theta_X, \Theta_Y} P(\theta_X, \theta_Y \mid G_0) P(D \mid \theta_X, \theta_Y, G_0) d \mid \theta_X, \theta_Y$$

Assuming that the parameters are independent, we have the following equation:

$$P(D|G_0) = \left(\int_{\Theta_X} P(\theta_Y \mid G_0) \prod_m P(x[m] \mid \theta_X, G_0) d\theta_X \right) \cdot$$

$$\left(\int_{\Theta_Y} P(\theta_Y \mid G_0) \prod_m P(y[m] \mid \theta_Y, G_0) d\theta_Y \right)$$

In the preceding equation, we can see that we have a marginal likelihood for each of the variables, X and Y. Now, if both of these variables are multinomial and have a Dirichlet prior, we can write each of these terms in the form of the equation that we discussed in the preceding section.

Now, let's consider the case of G_1. Again, assuming parameter independence, we can decompose the integral as follows:

$$P(D \mid G_1) = \left(\int_{\Theta_X} P(\theta_X \mid G_1) \prod_m P(x[m] \mid \theta_X, G_1) d\theta_X \right) \cdot$$

$$\left(\int_{\Theta_{Y \mid x^0}} P(\theta_{Y \mid x^0} \mid G_1) \prod_{m\,:x[m]=x^0} P(y[m] \mid \theta_{Y \mid x^0}, G_1) d\theta_{Y \mid x^0} \right) \cdot$$

$$\left(\int_{\Theta_{Y \mid x^1}} P(\theta_{Y \mid x^0} \mid G_1) \prod_{m\,:x[m]=x^1} P(y[m] \mid \theta_{Y \mid x^1}, G_1) d\theta_{Y \mid x^1} \right)$$

Now, let's compare the marginal likelihood of both the cases. If we choose the priors, $P(\Theta_X \mid G_0)$ and $P(\Theta_X \mid G_1)$, to be the same in both cases, we can see that the first terms in both cases are equal. Thus, given the assumption about the prior, the difference in the marginal likelihood is due to the difference in the marginal likelihood in all the observations over X, and all the observations over Y when we split the observations based on the value of X. Therefore, if Y has a different distribution in the split using observations of X, the latter term will have a better marginal likelihood. If the distribution is same in both splits, the simpler model will have a higher marginal likelihood. Thus, we can solve the problem that we faced in the case of maximum likelihood scoring.

Let's take an example of a learning structure using pgmpy:

```
In [1]: import numpy as np
In [2]: import pandas as pd
In [3]: from pgmpy.models import BayesianModel
In [4]: from pgmpy.estimators import BayesianEstimator

# Generating random data for two coin tossing examples
In [4]: raw_data = np.random.randint(low=0, high=2,
                                     size=(1000, 2))
In [5]: data = pd.DataFrame(raw_data, columns=['X', 'Y'])
In [6]: print(data)
Out[6]:
     X  Y
0    0  1
1    1  0
2    1  1
3    1  1
4    1  1
```

```
5      0    0
..     ..   ..
995    0    0
996    1    1
997    1    0
998    0    0
999    0    0

[1000 rows x 2 columns]
In [7]: coin_model = BayesianModel()
In [8]: coin_model.fit(data, estimator=BayesianEstimator)
In [9]: coin_model.get_cpds()
Out[9]:
[<TabularCPD representing P(X: 2) at 0x7f57bda018d0>,
 <TabularCPD representing P(Y: 2) at 0x7f57bda0124a>]

In [10]:coin_model.nodes()
Out[10]:['X', 'Y']

In [11]: coin_model.edges()
Out[11]: []
```

Let's take another example for the late-for-school model:

```
In [12]: raw_data = np.random.randint(low=0, high=2,

                                    size=(1000, 6)
In [12]: data = pd.DataFrame(raw_data, columns=['A', 'R', 'J',

                                                'G', 'L', 'Q'])

In [13]: student_model = BayesianModel()
In [14]: student_model.fit(data, esitmator=BayesianEstimator)

In [15]: student_model.get_cpds()
Out[15]:
[<TabularCPD representing P(A: 2) at 0x7a57e462d128>,
 <TabularCPD representing P(R: 2) at 0x7c57ad993048>,
 <TabularCPD representing P(J: 2) at 0x7f17cd991160>,
 <TabularCPD representing P(G: 2) at 0x7e67b129a278>,
 <TabularCPD representing P(L: 2) at 0x7e37e4695390>,
 <TabularCPD representing P(Q: 2) at 0x7f67a289d649>]

In [16]:student_model.get_nodes()
Out[16]:[ 'A', 'R', 'J', 'G', 'L', 'Q']

In [17]:student_model.get_edges()
Out[17]:[]
```

As we had generated the data randomly, all the variables are independent.

Summary

In previous chapters, we considered that we know the structure of the network, which is not true in most of real-life cases. In such cases, we need to learn the structures from the data. In this chapter, we discussed the problem of learning the parameters and structures using just data samples. Firstly, we discussed two different techniques of parameter estimation, maximum likelihood estimation, and Bayesian estimation. We saw that in cases when the data samples given to us don't represent the underlying distribution, the Maximum Likelihood estimate fails to generalize over new data points. Then, we discussed the problem of learning the structure from the data using the same two techniques, that is, maximum likelihood and Bayesian learning. We showed that in the case of structure learning as well, maximum likelihood overfits the training data if we don't have enough samples.

In the next chapter, we will discuss the parameters and structures of Markov networks using data samples.

6
Model Learning – Parameter Estimation in Markov Networks

In the preceding chapter, we learned about parameters and structures from the data in the case of Bayesian networks. In this chapter, we will focus on learning parameters and structures in the case of Markov networks. As it turns out, the learning task in the case of Markov networks is more difficult. This is because of the partition function that comes in the probability distribution. Because this partition function depends on all factors, it doesn't let us decompose our optimization functions into separate terms, as in the case of Bayesian networks. Therefore, we have to use some iterative method over the optimization function to find the optimal point in the parameter space.

In this chapter, we will discuss the following topics:

- Maximum likelihood parameter estimation
- Learning with approximate inference
- Structure learning

Maximum likelihood parameter estimation

As in the case of Bayesian networks, we can also estimate the parameters in the case of Markov networks using maximum likelihood. Let's see in detail how maximum likelihood works in the case of Markov networks.

Likelihood function

Let's take a very simple example of the network, $X - Y - Z$. We have two potentials, $\phi_1(X,Y)$ and $\phi_2(Y,Z)$. We can now define the joint distribution over this network as follows:

$$P(X,Y,Z) = \frac{1}{Z} \cdot \phi_1(X,Y) \cdot \phi_2(Y,Z)$$

Here, Z is the partition function and is defined as follows:

$$Z = \sum_{x,y,z} \phi_1(X,Y) \cdot \phi_2(Y,Z)$$

Therefore, the log-likelihood equation for a single instance $<x, y, z>$ would be as follows:

$$\ln P(x,y,z) = \ln \phi_1(x,y) + \ln \phi_2(y,z) - \ln Z$$

Suppose we have a dataset D containing M samples, we can write the likelihood in the following way:

$$P(\theta:D) = \prod_m \frac{1}{Z(\theta)} \cdot \phi_1(x[m],y[m]) \cdot \phi_2(y[m],z[m])$$

Thus, the log-likelihood equation translates to the following formula:

$$\ln P(\theta:D) = \sum_m \left(\ln \phi_1(x[m],y[m]) + \ln \phi_2(y[m],z[m]) - \ln Z \right)$$

$$= \sum_{x,y} M[x,y] \ln \phi_1(x,y) + \sum_{y,z} M[y,z] \ln \phi_2(y,z) - M \ln Z(\theta)$$

As we have seen in the case of Bayesian networks, once we have sufficient statistics that summarize the data (the joint count of the variables), we can learn the parameter, θ. However, with Markov models, the problem is the third term appearing in the earlier equation, that is, $M \ln Z(\theta)$:

$$Z(\theta) = \phi_1(x,y) \cdot \phi_1(y,z)$$

Thus, we get the following formula:

$$M \ln Z(\theta) = M\left(\ln \phi_1(x, y) + \ln \phi_2(y, z)\right)$$

So, the term *ln Z(Θ)* couples both ϕ_1 and ϕ_2. This poses a serious issue when we want to estimate the parameters by maximizing the likelihood. If we change the potential ϕ_1, it will change the value of ϕ_2 due to the coupling introduced by Z(Θ). So, unlike the Bayesian network, we cannot estimate the conditional probabilities independent of each other.

However, this problem can be solved for this specific network by treating the Markov model (*X − Y − Z*) as a Bayesian model *X→Y→Z*. Thus, learning the parameters of this Bayesian model, which are *P(X)*, *P(Y | X)*, and *P(Z | Y)*. Once we have estimated these parameters, it can be converted again to the context of a Markov model:

$$\phi_1(X,Y) = P(X) \cdot P(Y \mid X)$$

$$\phi_2 = P(Z \mid Y)$$

However, the caveat for this method is that not all Markov models can be converted into Bayesian models. For example, the diamond-shaped model represented in Fig 6.1 cannot be converted into a Bayesian model:

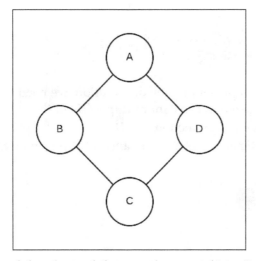

Fig 6.1: Diamond-shaped network that cannot be converted into a Bayesian model

Before we go into further discussion to learn the parameters for a Markov model, let's discuss a particular representation of it called as the log-linear model.

Log-linear model

A feature, $f(D)$, is a function from a subset of variables ranging from D to $\mathbb{R}$. It is similar to the factor with the non-negativity constraint. A special type of feature is the indicator feature. An indicator feature is such that it is 1 for some values of $y \in Val(\mathcal{D})$ and 0 otherwise.

Suppose that ϕ is a factor over a subset of variables represented by D. The factor, $\phi(\mathcal{D})$, can also be expressed as follows:

$$\phi(\mathcal{D}) = \exp\left(-\epsilon(\mathcal{D})\right)$$

Here, $\epsilon(D)$ is called as the energy function. Thus, the energy function is simply represented as follows:

$$\epsilon(D) = -\ln \phi(D)$$

Let's consider two random variables, X and Y, both of them have the cardinality, m. Let's assume their distribution is such that they are more favorable to situations when both of them have the same value. So, their energy function may be something like this:

$$\epsilon(X,Y) = \begin{cases} 10 & X = Y \\ 0 & otherwise \end{cases}$$

If we have a full factor representing the distribution, we need to have the m^2 values. This could be represented as a constant multiple (10) of the indicator feature for the event $X = Y$. Thus, the energy function, $\epsilon_i(D_i)$, can be compactly represented as $w_i \cdot f_i(\mathcal{D}_i)$ (in this particular case, w_i is 10 and $f_i(\mathcal{D}_i)$ is an identity feature).

From our previous discussion, we know that the joint probability distribution of random variables, $X_1, X_2, ..., X_n$, encoded by a Markov model will be the following formula:

$$P(X_1, X_2, ..., X_n) = \frac{1}{Z(\theta)} \prod_{i=1}^{k} \phi_i(\mathcal{D}_i)$$

$$= \frac{1}{Z(\theta)} \exp\left[-\sum_{i=1}^{k} \epsilon_i(\mathcal{D}_i)\right]$$

$$= \frac{1}{Z(\theta)} \exp\left[-\sum_{i=1}^{k} w_i \cdot f_i(\mathcal{D}_i)\right]$$

This type of representation of a Markov model is called a log-linear model. A log-linear model is associated with a set of features, $\{f_1(\mathcal{D}_1), f_2(\mathcal{D}_2), ..., f_k(\mathcal{D}_k)\}$, and weights, $\{w_1, w_2, ..., w_k\}$, where $\mathcal{D}_i$ is a complete sub-graph of the model and is expressed as follows:

$$P(X_1, X_2, ..., X_n) = \frac{1}{Z(\theta)} \exp\left[-\sum_{i=1}^{k} w_i \cdot f_i(\mathcal{D}_i)\right]$$

Let's go back to our previous discussion about estimating the parameters of a Markov model by maximizing the likelihood. We can write the probability distribution for $P(X_1, X_2, ..., X_n : \theta)$, as follows:

$$P(X_1, X_2, ..., X_n : \Theta) = \frac{1}{Z(\Theta)} \exp\left[-\sum_{i=1}^{k} \theta_i \cdot f_i(\mathcal{D}_i)\right]$$

So, the likelihood function for a dataset, D, containing M examples will be as follows:

$$P(\Theta : D) = \prod_{j=0}^{m} \frac{1}{Z(\Theta)} \exp\left[-\sum_{i=1}^{k} \theta_i \cdot f_i(\mathcal{D}_i(j))\right]$$

Thus, the log-likelihood equation will be as follows:

$$\ln P(\Theta : D) = \sum_{j=1}^{m} \left(\sum_{i=1}^{k} \theta_i \cdot f_i(D_i(j)) - \ln Z(\Theta) \right)$$

$$= \sum_{j=1}^{m} \left(\sum_{i=1}^{k} \theta_i \cdot f_i(D_i(j)) \right) - \sum_{j=1}^{m} \ln Z(\Theta)$$

$$= \sum_{i=1}^{k} \theta_i \cdot \left(\sum_{j=1}^{m} f_i(D_i(j)) \right) - M \ln Z(\Theta)$$

Dividing both sides of the preceding equation by M, we get the following formula:

$$\frac{1}{M} \ln P(\Theta : D) = \sum_{i=1}^{k} \theta_i \cdot \frac{1}{M} \left(\sum_{j=1}^{m} f_i(D_i(j)) \right) - \ln Z(\Theta)$$

$$= \sum_{i=1}^{k} \theta_i \cdot \mathbb{E}_D \left[f_i(D_i) \right] - \ln Z(\Theta)$$

Here, $\mathbb{E}_D \left[f_i(D_i) \right]$ is the empirical expectation of f_i, that is, its average over the dataset.

Gradient ascent

In the preceding section, we saw that to estimate the parameters, we have to maximize the earlier equation. As the equation is not in a closed form, we have to use some iterative techniques to compute the maximum values. One of the simplest iterative techniques is the gradient ascent. In this section, we will mainly focus on gradient ascents and how to use them.

In this method, we will start with a random point on the curve and move upward in the direction of the gradient. So, if $x^{(t)}$ is the point that we got in the previous iteration, then $x^{(t+1)}$ would be as follows:

$$x^{(t+1)} \leftarrow x^{(t)} + \eta \nabla f \left(x^{(t)} \right)$$

This is repeated until we can't go any further, that is, $\left\| x^{(t+1)} - x^{(t)} \right\| < \delta$, where δ is the convergence threshold. This is analogous to climbing up the hill. For example, let's consider a simple equation, such as $f(x) = sin\ x$. We want to find the maxima of the given function in the interval from 0 to π.

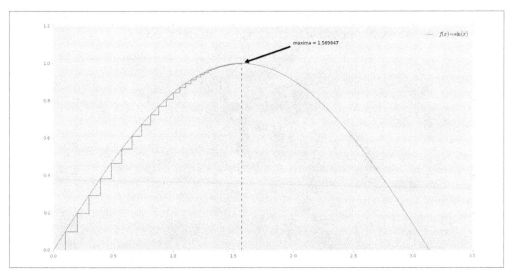

Fig 6.2: Plot showing the steps taken in each step of the gradient ascent

In Fig 6.2, the green curve shows the steps taken in each iteration of the gradient ascent. We can see that when we reach the maxima, the gradient approaches 0, thus x, with each iteration, the value of x will keep decreasing, that is, $\left\| x^{(t+1)} - x^{(t)} \right\| < \delta$.

The performance of the gradient ascent depends on the choice of η. If η is too large (as shown in Fig 6.3), we will overshoot the maxima in each iteration. If η is too small (as shown in Fig 6.4), we will require a lot of iterations to converge. In practice, the value of η should be adaptive. It should start with the large of η and reduce in each iteration:

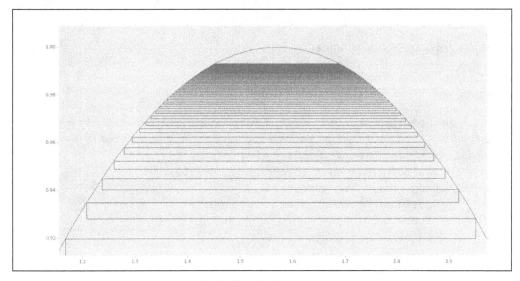

Fig 6.3: The n in this case is 2

In each iteration, we can see that the algorithm overshoots the maxima within 100 iterations, and that we are not able to converge with the maxima.

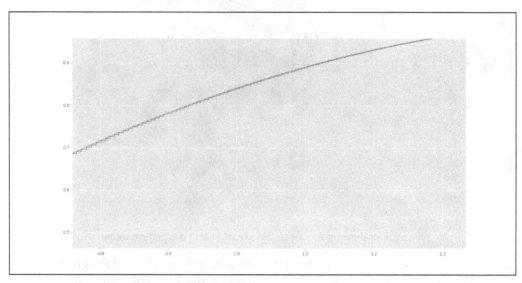

Fig 6.4: The *n* in this case is 0.01

The green curve shows the steps taken in each iteration. As the steps are very small, it takes a lot of iterations to converge. In this case, it was 758.

In real life, we generally don't use the gradient ascent algorithms. Instead, we use one variant of it called the conjugate gradient ascent. The conjugate gradient method solves the issue of overshooting by adding a friction term, that is, each step depends on the last two values of the gradient, and sharp turns are avoided.

In Python, this can be implemented as follows:

```
In [1]: import numpy as np
In [2]: from scipy import optimize

# The methods implemented in scipy are meant to find the minima,
# thus to find the maxima we have to negate the functions
In [3]: f_to_optimize = lambda x: -np.sin(x)
In [4]: optimize.fmin_cg(f_to_optimize, x0=[0])
Optimization terminated successfully.
         Current function value: -1.000000
         Iterations: 2
         Function evaluations: 15
         Gradient evaluations: 5
Out[5]: array([ 1.57079632])
```

There are multiple methods that use second-order methods for faster convergence, such as L-BGFS. The detailed descriptions of these algorithms are out of the scope of this book.

Let's go back to our discussion on the estimation of the parameters for a Markov model. In the previous section, we saw that the log-likelihood equation was as follows:

$$\frac{1}{M} l(\Theta : D) = \sum_{i=1}^{k} \theta_i \cdot \mathbb{E}_D\left[f_i(\mathcal{D}_i)\right] - \ln Z(\Theta)$$

Here, $l(\Theta : D)$ is the log-likelihood function and is defined as $\ln P(\Theta : D)$. For any gradient-based method, we need to have the gradient of the log-likelihood function with respect to θ_i. The gradient is computed as follows:

$$\frac{\partial}{\partial \theta_i} \frac{1}{M} l(\Theta : D) = \mathbb{E}_D\left[f_i(\mathcal{D}_i)\right] - \mathbb{E}_\Theta\left[f_i\right]$$

As we know, at the maxima, the value of the derivate would be 0, thus at the maxima, the value of $\mathbb{E}_D\left[f_i(\mathcal{D}_i)\right] = \mathbb{E}_\Theta\left[f_i\right]$ (the expected value of each feature relative to the distribution $\mathbb{E}_\Theta\left[f_i\right]$) matches its empirical expectation, $\mathbb{E}_D\left[f_i(\mathcal{D}_i)\right]$, in D. We discussed earlier that to compute the value of Θ, we have to retort to an iterative method. For the iterative method, we need to compute the gradient. From the preceding equation, we know that the gradient is the difference between the empirical expectation of the feature in the data (its empirical count) and its expected count relative to the current parameterization. For example, let's consider a Markov model, as given in Fig 6.5, between the two binary-valued random variables, A and B:

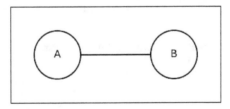

Fig 6.5: Markov model representing the dependencies among two
random variables, A and B, each of them being binary valued

Considering that the features used for the model are only indicator features, that is, $f_{a_0,b_0}(a,b)$, $f_{a_1,b_0}(a,b)$, $f_{a_0,b_1}(a,b)$, and $f_{a_1,b_1}(a,b)$, where $f_{a_0,b_0}(a,b)$ is simply $I(a=a_0) \cdot I(b=b_0)$. $I(a=a_0)$ being the indicator function would be 1 if $a=a_0$; otherwise, it would be 0. Then, $\mathbb{E}_D\left[f_{a_0,b_0}\right]$ will be the empirical frequency of a_0,b_0 in the dataset D, and $\mathbb{E}_\Theta\left[f_{a_0,b_0}\right]$ will be the probability of getting a_0,b_0 for a particular value of Θ, that is, $\mathrm{E}_\Theta[a_0,b_0]$. So, the gradient would be the difference between the two numbers, as stated earlier.

However, this method poses a serious issue. To compute the gradient at each step, we need to compute the value of $\mathrm{E}_\Theta[a_0,b_0]$. This requires the inference to be run over the whole network. Thus, in each iteration of the gradient ascent step, we need to run the inference over the complete network, which is computationally very expensive and also intractable sometimes.

Now, let's see some code examples to learn parameters using pgmpy:

```
In [1]: import numpy as np
In [2]: import pandas as pd
In [3]: from pgmpy.models import MarkovModel
In [4]: from pgmpy.estimators import MaximumLikelihoodEstimator

# Generating some random data
In [5]: raw_data = np.random.randint(low=0, high=2, size=(100, 2))
In [6]: raw_data
Out[6]:
array([[1, 1],
       [1, 1],
       [0, 1],
       ...,
       [0, 0],
       [0, 0],
       [0, 0]])
In [7]: data = pd.DataFrame(raw_data, columns=['A', 'B'])
In [8]: data
Out[8]:
      A  B
0     1  1
1     1  1
2     0  1
3     1  0
..    .. ..
996   1  1
997   0  0
998   0  0
999   0  0
```

```
[1000 rows x 2 columns]

# Markov Model as stated in Fig 6.5
In [9]: markov_model = MarkovModel([('A', 'B')])
In [10]: markov_model.fit(data,
                          estimator=MaximumLikelihoodEstimator)
In [11]: factors = coin_model.get_factors()
In [12]: print(factors[0])
Out[12]:
```

A	B	phi(A,B)
A_0	B_0	0.1000
A_0	B_1	0.2000
A_1	B_0	0.4600
A_1	B_1	0.2400

Learning with approximate inference

In the preceding section, we saw that to estimate parameters using the maximum likelihood method, we need to run the inference algorithm in each step or iteration of the learning method to compute $\mathbb{E}_\Theta[f_i]$. This is irrespective of the learning method that we use. Running the exact inference over the whole network is computationally expensive, and sometimes, intractable. For example, in the case of a grid network, the exact inference algorithms are computationally intractable.

There are multiple ways to overcome this issue. One way is to treat the inference algorithm as a black box independent of the learning algorithm. This approach has its own advantages and disadvantages. It allows us to use approximate inference algorithms instead of exact ones, which are computationally tractable. However, at the same time, the inaccuracy in computing the gradient might lead to oscillations in the learning algorithm, thus affecting its convergence.

Another way of solving this issue is to use the alternative approximate objective functions whose optimization does not require the inference to be run over the whole network. Unlike the previous method, which approximately optimized the likelihood function, this method optimizes an approximated likelihood function exactly. Although it may seem that both of these methods try to do the same thing, one way or the other, the second approach is more useful, as it allows us to use any applicable optimization algorithm and also allows us to bind the error in the optimum values.

Belief propagation and pseudo-moment matching

One of the most popular inference algorithms is the belief propagation. One way to use the belief propagation is to simply run it in each step of a gradient ascent to compute the expected value of a feature with respect to the distribution. In the previous chapters, we studied the family preservation property of the cluster graph. Due to the family preservation property, we can say that each feature, f_i, will be a subset of a cluster, C_i, in the cluster graph. Thus, to compute the value of $\mathbb{E}_\Theta[f_i]$, we can simply compute the belief propagation marginals of C_i, and then compute its expectation. However, this approach has some serious issues. In the case of grid graphs, the belief propagation won't converge. Thus, the gradient computed using the marginals will be oscillatory. So, any gradient-based optimization algorithm won't converge.

One solution to this issue is to use the convergent version of the belief propagation. For example, the belief propagation using approximate messages or using an alternate objective function.

As we discussed in the preceding section, at convergence, $\mathbb{E}_D[f_i(\mathcal{D}_i)] = \mathbb{E}_\Theta[f_i]$, which is the expected value of each feature relative to the distribution, ($\mathbb{E}_\Theta[f_i]$) matches its empirical expectation in D ($\mathbb{E}_D[f_i(\mathcal{D}_i)]$). This can be reformulated in the context of the belief propagation as follows:

$$\mathbb{E}_D[f_i(\mathcal{D}_i)] = \mathbb{E}_{\beta_i(C_i)}[f_i]$$

The distribution of the cluster graph is parameterized by the set of the cluster potentials, ($\beta_i(C_i)$). Let's assume that for each cluster, C_i, in the cluster graph and for each assignment of C_i^j in the cluster C_i, we have an indicator feature, $I(C_i^j)$ (which is 1 when $C_i = c_i^j$; otherwise, it will be 0). Thus, the preceding equation, $\mathbb{E}_D[f_i(\mathcal{D}_i)] = \mathbb{E}_{\beta_i(C_i)}[f_i]$, translates to the following formula:

$$\beta_i(c_i^j) = P(c_i^j)$$

At the convergence of the gradient ascent algorithm, the belief of cluster C_i must be the same as the empirical frequencies of the variables present in C_i in the data. This formulation gives us a major advantage — if we already know the outcome of the convergence, there is no need to run the algorithm.

As the full table parameterization of a Markov model is redundant, we can have multiple solutions that give rise to the same beliefs. One such solution could be obtained by dividing the cluster potential of a particular cluster, C_i (β_i) by the sepset potential, $\mu_{i,j}$. This can be described as follows:

$$\phi_i \leftarrow \frac{\beta_i}{\mu_{i,j}}$$

This method can be summarized as follows:

1. For each cluster, C_i, compute the cluster potential, β_i, as the empirical frequencies of the variables present in the cluster from the data, that is, $\beta_i\left(c_i^j\right) = P\left(c_i^j\right)$.
2. Run a single pass of a message passing algorithm to calibrate the graph. Compute the sepset potential, $\mu_{i,j}$, corresponding to a sepset, $S_{i,j}$, between C_i and C_j.

3. Compute the final factors as $\phi_i \leftarrow \dfrac{\beta_i}{\mu_{i,j}}$.

Now, let's see some code examples for how to learn parameters using pgmpy:

```
In [1]: import numpy as np
In [2]: import pandas as pd
In [3]: from pgmpy.models import MarkovModel
In [4]: from pgmpy.estimators import PseudoMomentMatchingEstimator

# Generating some random data
In [5]: raw_data = np.random.randint(low=0, high=2, size=(100, 4))
In [6]: raw_data
Out[6]:
array([[1, 1, 0, 0],
       [1, 1, 1, 0],
       [0, 1, 0, 1],
       ...,
       [0, 0, 0, 0],
       [0, 0, 0, 0],
       [0, 0, 1, 1]])
In [7]: data = pd.DataFrame(raw_data, columns=['A', 'B', 'C', 'D'])
In [8]: data
Out[8]:
     A  B  C  D
0    1  1  0  0
1    1  1  1  0
2    0  1  0  1
```

```
3     1   0   0   0
..    ..  ..  ..  ..
996   1   1   0   1
997   0   0   0   0
998   0   0   0   0
999   0   0   1   1

[1000 rows x 4 columns]

# Diamond shaped Markov Model as stated in Fig 6.1
In [9]: markov_model = MarkovModel([('A', 'B'), ('B', 'C'),
                                     ('C', 'D'), ('D', 'A')])
In [10]: markov_model.fit(data,
                          estimator=PseudoMomentMatchingEstimator)
In [11]: factors = coin_model.get_factors()
In [12]: factors
Out[12]:
[<Factor representing phi(A:2, B:2) at 0x7f244d0f5e87>,
 <Factor representing phi(B:2, C:2) at 0x7f244d0f5e97>,
 <Factor representing phi(C:2, D:2) at 0x7f244d0f5f10>,
 <Factor representing phi(D:2, A:2) at 0x7f244d0f5f24>]
```

Structure learning

In the preceding chapter, we discussed structure learning in the case of Bayesian models. In this section, we will focus on structure learning in the case of Markov models. Similar to structure learning in the case of Bayesian models, here, we are also going to focus on the two methods of structure learning. The first one being a constraint-based approach, which tries to search for a graph structure, satisfying the independence conditions observed from the data. The other approach is score-based in which we define an objective function for a different model, and then search for a high-scoring model.

Constraint-based structure learning

In the preceding chapter, we discussed the constraint-based structure learning in the case of a Bayesian model. In Markov models, this approach seems to be more advantageous as compared to the scoring-based approach. As the independence conditions for Markov models are much simpler than those in Bayesian models (which involve d-separation), the algorithms inferring the structure are much simpler. The other major advantage is that the scoring-based structure learning uses the likelihood function. From our previous discussion, we know that computing the likelihood is computationally expensive in the case of Markov models, and in some cases, it may be intractable as well.

On the other hand, the constraint-based approaches have some disadvantages as well. As we try to find dependence conditions among variables from the data, this method is not robust to the noise present in the data. So, if we get a wrong result from our dependence tests, our whole learning fails. Secondly, these methods only learn the structure of the model, not the distribution. To obtain the distribution, we must use the methods that we have for parameter estimation (which we had discussed in the preceding section).

Before going into a detailed discussion about constraint-based learning, let's first recapitulate the independencies of a Markov model:

- **Local Markov independencies**: This independence is of a variable, X, from the rest of the variables in the model given its Markov blanket, that is, $\left(X \perp \mathcal{X} - \{X\} - MB_{\mathcal{H}}(X) \mid MB_{\mathcal{H}}(X) \right)$
- **Pair-wise independencies**: This independence is of each nonadjacent pair of variables, X, Y, given all the other variables, $\left(X \perp Y \mid \mathcal{X} - \{X, Y\} \right)$
- **Global independencies**: This independence includes all the independencies present due to the separation among the variables in the graph

Let's go back to our discussion on structure learning. Assume that we have a distribution, P^*, that can be represented by a Markov model, $\mathcal{H}^*$, so that $\mathcal{H}^*$ is a perfect map for P^*. Our objective is to find $\mathcal{H}^*$ by performing the earlier-stated independence tests on P^*. However, none of the independencies can be checked tractably, as they all involve the entire set of variables, $\mathcal{X}$. Apart from being computationally intractable, this also poses some serious statistical issues. One of them is that the independence assertions are evaluated on the empirical data and not on the true distribution. Secondly, to estimate the distribution sufficiently well, we need many data points exponentially.

To overcome this issue, we need to come up with an alternative set of independencies that involves only small subsets of variables. For example, if in a network, $\mathcal{H}^*$, X, and Y are not neighbors, they are separated by the Markov blankets, $MB_{\mathcal{H}^*}(X)$ and $MB_{\mathcal{H}^*}(Y)$. Thus, we can find a set, Z, such that $|Z| \leq \min\left(\left| MB_{\mathcal{H}^*}(X) \right|, \left| MB_{\mathcal{H}^*}(X) \right| \right)$. On the other hand, if X and Y are neighbors in $\mathcal{H}^*$, we cannot find any such set, Z. Thus, we can state the following equation:

$$X - Y \notin \mathcal{H}^* \text{ if and only if } \exists Z, |z| \leq d^* \& P^* \models \left(X \perp Y \mid Z \right)$$

Here, $d*$ is the maximum cardinality of any variable. Thus, we can see that to determine whether $X - Y$ is present in $\mathcal{H}*$, we need to run $\sum_{k=0}^{d*}\binom{n-2}{k}$ independence tests with tests involving only $d*+2$ variables. So, for small values of $d*$, it is computationally tractable.

Although these algorithms work fairly in some cases, they have some very fundamental limitations:

- The number of samples required to obtain correct results for all the independent tests are too high.

- This algorithm assumes that there is a Markov model, $\mathcal{H}*$ present, which is a perfect map of the distribution, $P*$. At the most, the cardinality of a variable can only be $d*$. The violation of any of these assumptions will lead to learning incorrect network structures.

Score-based structure learning

In the preceding chapter, we learned that in score-based structure learning, we define a hypothesis space consisting of possible networks and an objective function, which is required to score different networks, and then we construct a search algorithm that attempts to find the network structure with the highest score in the hypothesis space. In the case of Markov models, we will be following similar principles.

Let's first discuss the formulation of the hypothesis space. There are many ways of formulating the hypothesis space, depending on the granularity at which they consider network parameterization:

- The coarsest-grained hypothesis space being the space of different structures of the Markov model.

- At the next level, we could consider the network parameterization to be the size of the factors in the graph. In this case, the hypothesis space is a space consisting different factor graphs.

- At the finest level of granularity, we can consider the hypothesis space to be at the level of individual features in a log-linear model and measure the sparcity at the level of the features included in the model.

As the level of granularity of the hypothesis space increases, it allows us to select a parameterization that matches the properties of the distribution of data without over-fitting. For example, the hypothesis space at the granularity of a factor graph allows us to distinguish between a single large factor over k variables and a set of $\binom{k}{2}$ pair-wise factors over the same variables (which require far less parameters). However, at the same time, finer-grained spaces can obscure the connection to network structures. For example, if we are dealing with the hypothesis space at the level of individual features, the addition of a single feature, $f(d)$, into the model will increase the complexity of the model by introducing edges between all the variables present in d. This will create an issue while performing an inference in the model.

In this section, we will focus on score-based structure learning, with the hypothesis space being at the level of individual features. Considering our hypothesis space to be Ω, our task is to select a log-linear model structure, $\mathcal{M}$, which is defined by a subset of features, $\Phi[\mathcal{M}] \subseteq \Omega$. Assume that $\Phi[\mathcal{M}]$ is the set of parameters, θ, that are compatible with the model structure. This can also be written as $\theta_i \neq 0$ only if $f_i \in \Phi[\mathcal{M}]$. The structure and compatible parameterization define a log-linear model distribution as follows:

$$P(\mathcal{X} \mid \mathcal{M}, \theta) = \frac{1}{Z} \exp \left\{ \sum_{i \in \Phi[\mathcal{M}]} \theta_i \cdot f_i(\xi) \right\}$$

$$= \frac{1}{Z} \exp \left\{ f^T \theta \right\}$$

Sometimes, in addition to the objective function, we also want to impose some additional structural constraints. For example, we may want to bind the tree width of the graph structure. These constraints help in rejecting very dense networks, thus reducing the chances of over-fitting.

The likelihood score

Similar to the likelihood score discussed for Bayesian models, the likelihood score for Markov models is defined as follows:

$$score_L = \max_{\theta \in \Theta[\mathcal{M}]} \ln P(D \mid \mathcal{M}, \theta)$$

$$= l\left(\left\langle \mathcal{M}, \hat{\theta}_{\mathcal{M}} \right\rangle : D \right)$$

Here, $\hat{\theta}_M$ are the maximum likelihood parameters compatible with $\mathcal{M}$. Here, as well, the likelihood score measures the fitness of the model to the data. Further, in this case, the likelihood score tries to select a more complex model as it could capture the noise in the data very well. So, in reality, the likelihood scores are used only with very strict constraints on the structure of the model. For example, putting an upper bound on the tree width of the graph structure.

Let's see an example of tossing two binary variable models using pgmpy:

```
In [1]: import numpy as np
In [2]: import pandas as pd
In [3]: from pgmpy.models import MarkovModel
In [4]: from pgmpy.estimators import MaximumLikelihoodEstimator

# Generating random data
In [4]: raw_data = np.random.randint(low=0, high=2,
                                     size=(1000, 2))
In [5]: data = pd.DataFrame(raw_data, columns=['X', 'Y'])
In [6]: model = MarkovModel()
In [7]: model.fit(data, estimator=MaximumLikelihoodEstimator)

In [8]: model.get_factors()
Out[8]:
[<Factor representing phi(X:2, Y:2) at 0x7f244d0f5e87>]
In [9]: model.nodes()
Out[9]:
['X', 'Y']
In [10]:coin_model.edges()
Out[10]:
[('X', 'Y')]
```

Bayesian score

In the preceding chapter, we discussed the Bayesian score whose primary term is a marginal likelihood that integrates the likelihood over all the possible network parameterizations, that is, $\int P(D\,|\,\mathcal{M},\theta)P(\theta\,|\,\mathcal{M})\,d\theta$. This avoided over fitting by preventing overly optimistic or complex models from fitting into the training data. However, unlike Bayesian models, in the case of Markov models, it is not as easy to evaluate the likelihood. Thus, evaluating the marginal likelihood becomes a challenge in this case.

So, in this case, we use asymptotic approximation of the marginal likelihood:

$$score_{BIC} = l\left(\left\langle \mathcal{M}, \hat{\theta}_{\mathcal{M}} \right\rangle : D\right) - \frac{\dim(\mathcal{M})}{2} \ln M$$

Here, $\dim(\mathcal{M})$ is the dimension of the model and M is the number of instances of the dataset, D. It measures the degree of freedom of our parameter space. When the model has non-redundant features, $\dim(\mathcal{M})$ is exactly the number of features. Also, when we have redundant features, it is less than the number of features.

Let's see an example of tossing two binary variable models using pgmpy:

```
In [1]: import numpy as np
In [2]: import pandas as pd
In [3]: from pgmpy.models import MarkovModel
In [4]: from pgmpy.estimators import BayesianEstimator

# Generating random data
In [4]: raw_data = np.random.randint(low=0, high=2,
                                     size=(1000, 2))
In [5]: data = pd.DataFrame(raw_data, columns=['X', 'Y'])
In [6]: model = MarkovModel()
In [7]: model.fit(data, estimator=BayesianEstimator)

In [8]: model.get_factors()
Out[8]:
[<Factor representing phi(X:2, Y:2) at 0x7f244d0f5e87>]
In [9]: model.get_nodes()
Out[9]:
['X', 'Y']
In [10]:coin_model.get_edges()
Out[10]:
[('X', 'Y')]
```

Summary

In the previous chapters, we discussed learning the parameters, as well as the structures, of a Bayesian model using just the data samples. In this chapter, we discussed the same situations, but in the context of a Markov model. Firstly, we discussed a very famous technique of parameter estimation, maximum likelihood estimation. We saw that in Markov models, even the maximum likelihood estimate in the case of a simple model could be computationally expensive, and in some cases, it could also be intractable. This motivated us to find alternatives, such as using approximate inference algorithms to compute the gradient or using a different likelihood. We showed that learning with belief propagation can be reformulated as optimizing inference and learning simultaneously. Then, we discussed the problem of learning the structure from the data using the same two techniques, maximum likelihood and Bayesian learning.

In the next chapter, we will discuss some of the most commonly used special cases of the Bayesian and Markov networks, such as Naive Bayes and dynamic Bayesian networks.

7
Specialized Models

In the previous chapters, we discussed the generic cases of models. Now we have a good understanding of these models. In this chapter, we will discuss some of the special cases of Bayesian and Markov networks that are extensively used in real-life problems.

In this chapter, we will be discussing:

- The Naive Bayes model
- Dynamic Bayesian networks
- The Hidden Markov model

The Naive Bayes model

The Naive Bayes model is one of the most efficient and effective learning algorithms, particularly in the field of text classification. Although over-simplistic, this model has worked out quite well. In this section, we are going to discuss the following topics:

- What is a Naive Bayes model?
- Why does it even work?
- Types of Naive Bayes models

Before discussing the Naive Bayes model, let's first discuss about the Bayesian classifier. A Bayesian classifier is a probabilistic classifier that uses the Bayes theorem to predict a class. Let c be a class and $X = \{x_1, x_2, ..., x_n\}$ be a set of features. Then, the probability of the features belonging to class c, that is $P(c \mid X)$, can be computed using the Bayes theorem as follows:

$$P(c \mid X) = \frac{P(c) \cdot P(X \mid c)}{P(X)}$$

So, for a given set of features, the output class can be predicted as follows:

$$\hat{c} = \arg\max_{c \in C} P(c \mid X)$$

$$= \arg\max_{c \in C} \frac{P(c) \cdot P(X \mid c)}{P(X)}$$

$$= \arg\max_{c \in C} P(c) \cdot P(X \mid c)$$

Here, $P(c)$ is the prior probability of the class c and $P(X \mid c)$ is the likelihood of X given c. If X were an univariate feature, then computing $P(X \mid c)$ would be $P(x_1 \mid c)$, which is easier to compute. However, in the case of multivariate features, $P(X \mid c)$ is as follows:

$$P(X \mid c) = P(x_1 \mid c) \prod_{i=2}^{n} P(x_i \mid x_{i-1}, ..., x_1, c)$$

The Naive Bayes model simplifies the computation of $P(X \mid c)$ by taking a strong independence assumption over the features.

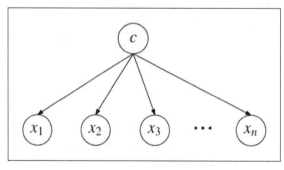

Fig 7.1: Graphical model representing the Naive Bayes model

Fig 7.1 shows the graphical model corresponding to the naive assumption of a strong independence among the features. It assumes that any features x_i and x_j are conditionally independent of each other given their parent c (or $x_i \perp x_j \mid c \, \forall i, j \epsilon [1, n]$). Thus, $P(X \mid c)$ can be stated as follows:

$$P(X \mid c) = \prod_{i=1}^{n} P(x_i \mid c)$$

For example, let's say we want to classify whether a given ball is a tennis ball or a football, and the variables given to us are the diameter of the ball, the color of the ball, and the type of surface. Here, the color of the ball, the size, and surface type are clearly independent variables and the type of ball depends on these three variables giving a network structure, as shown in Fig 7.2:

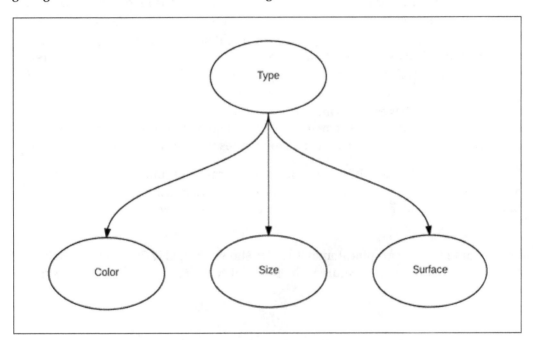

Fig 7.2: Network structure for the ball classification example

Why does it even work?

Although over-simplistic in assumption regarding dependence between the features, the Naive Bayes algorithm has performed very well in some cases. Surprisingly, it also performs very well in cases where there exists a strong dependence between the features or attributes. In this section, we are going to unravel the mystery.

Let's start with a simple binary classification problem where we have to predict the output class c based on the feature X. As it is a binary classification, there are only two output classes. For the sake of simplicity, let's assume one class to be a positive class, represented as c^+ , and the other to be a negative class, represented as c^- . One simple explanation is that Naive Bayes owes its good performance to the zero-one loss function that defines the error as the number of incorrect classifications. Unlike other loss functions, such as squared error, the zero-one loss function does not penalize the incorrectness in estimating the probability as long as the maximum probability is assigned to the correct class. For example, for a given set of features X, the actual posterior probability $P\left(c^+ \mid X\right)$ might be 0.8, and $P\left(c^- \mid X\right)$ might be 0.2, but due to the naive assumption regarding the dependencies between features, Naive Bayes may predict $P\left(c^+ \mid X\right)$ as 0.6 and $P\left(c^- \mid X\right)$ as 0.4. Although the probability estimations are incorrect, the class predicted is same in both of these cases. Thus, Naive Bayes performed well in the case of strong dependencies between features. However, the fundamental question has not yet been answered: why couldn't the strong dependencies between features flip the classification?

Before discussing the details, let's introduce the formal definition of the equivalence of two classifiers under the zero-one loss function. Two classifiers f_1 and f_2 are said to be equal under the zero-one loss function, if for every X in the example space, $f_1(X) \geq 0$ and $f_2(X) \geq 0$. This is denoted as $f_1 = f_2$.

Let's assume the true graphical model $\mathcal{G}_T$ (as shown in the Fig 7.3) represents the dependencies between the features. The probability $P_{\mathcal{G}_T}\left(x_1, x_2, ..., x_n, c\right)$ can be stated as follows:

$$P_{\mathcal{G}_T}\left(x_1, x_2, ..., x_n, c\right) = P(c) \prod_{i=1}^{n} P\left(x_i \mid Pa_{\mathcal{G}_T}\left(x_i\right), c\right)$$

Here, $Pa_{\mathcal{G}_T}(x_i)$ represents the parent of x_i in $\mathcal{G}_T$, except for the class node c.

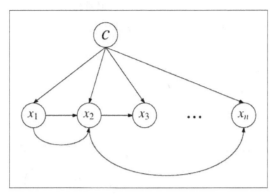

Fig 7.3: Graphical model representing strong dependencies between the features

To measure how strong the dependency between two features is, we have to quantify it. Naturally, the ratio of conditional probability of a node given its parents ($P\left(x_i \mid Pa_{\mathcal{G}_T}(x_i), c\right)$) over the conditional probability of the node without its parents ($P(x_i, c)$) reflects how strong the parent affects the node in each class. This parameter is called a local dependence derivate and is represented as follows:

$$dd^+_{\mathcal{G}_T}\left(x \mid Pa_{\mathcal{G}_T}(x)\right) = \frac{P\left(x_i \mid Pa_{\mathcal{G}_T}(x_i), c^+\right)}{P\left(x_i \mid, c^+\right)}$$

$$dd^-_{\mathcal{G}_T}\left(x \mid Pa_{\mathcal{G}_T}(x)\right) = \frac{P\left(x_i \mid Pa_{\mathcal{G}_T}(x_i), c^-\right)}{P\left(x_i \mid, c^-\right)}$$

When x has no parents, $dd^+_{\mathcal{G}_T}\left(x \mid Pa_{\mathcal{G}_T}(x)\right)$ is defined as 1. When $dd^+_{\mathcal{G}_T}\left(x \mid Pa_{\mathcal{G}_T}(x)\right) \geq 1$, it means x's local dependency supports class c^+, else it supports the class c^-. Similarly, $dd^-_{\mathcal{G}_T}\left(x \mid Pa_{\mathcal{G}_T}(x)\right) \geq 1$. This means x's local dependency supports class is c^-, else it supports class c^+. In the case where the local dependence derivate for each class supports the other, it means they partially cancel each other out and the final classification is the class with the maximum local dependence derivate. So, to check which class supports the local dependencies, we can use the ratio of the local dependence derivate of the two classes, represented as $ddr_{\mathcal{G}_T}$:

$$ddr_{\mathcal{G}_T} = \frac{dd^+_{\mathcal{G}_T}\left(x \mid Pa_{\mathcal{G}_T}(x)\right)}{dd^-_{\mathcal{G}_T}\left(x \mid Pa_{\mathcal{G}_T}(x)\right)}$$

If $ddr_{\mathcal{G}_T} > 1$, then x's local dependency supports class $ddr_{\mathcal{G}_T} < 1$. If c^-, then x's local dependency supports class $ddr_{\mathcal{G}_T} = 1$. If , the local dependence distributes evenly in both c^+ and c^-. Thus, the dependency does not affect the classification, however strong it may be.

As stated earlier, for classification using the Bayesian classifier, we use $\hat{c} = \arg\max_{c \in C} P(c) \cdot P(X \mid c)$. Thus, in cases of binary classification, we can define a variable $f(x_1, x_2, \dots, x_n, c)$ as the ratio of $P(c^+) \cdot P(X \mid c^+)$ over $P(c^-) \cdot P(X \mid c^-)$:

$$f(x_1, x_2, \dots, x_n, c) = \frac{P(c^+) \cdot P(X \mid c^+)}{P(c^-) \cdot P(X \mid c^-)}$$

If $f(x_1, x_2, \dots, x_n, c) \geq 1$, then the example is classified as belonging to the class c^+, else c^-. Summarizing all the earlier formulations, $f_{\mathcal{G}_T}(x_1, x_2, \dots, x_n, c)$ can be stated as follows:

$$f_{\mathcal{G}_T}(x_1, x_2, \dots, x_n, c) = \frac{P(c^+) \prod_{i=1}^{n} P(x^i \mid Pa_{\mathcal{G}_T}(x_i), c^+)}{P(c^-) \prod_{i=1}^{n} P(x^i \mid Pa_{\mathcal{G}_T}(x_i), c^-)}$$

$$= \left[\frac{P(c^+) \prod_{i=1}^{n} P(x^i \mid Pa_{\mathcal{G}_T}(x_i), c^+) P(x_i \mid c^-)}{P(c^-) \prod_{i=1}^{n} P(x^i \mid Pa_{\mathcal{G}_T}(x_i), c^-) P(x_i \mid c^+)} \right] \cdot \left[\prod_{i=1}^{n} \frac{P(x_i \mid c^+)}{P(x_i \mid c^-)} \right]$$

$$= \left[\prod_{i=1}^{n} ddr_{\mathcal{G}_T}(x_i) \right] \cdot f_{\mathcal{G}_{NB}}(x_1, x_2, \dots, x_n, c)$$

Here, $f_{\mathcal{G}_{NB}}(x_1, x_2, \dots, x_n, c)$ represents $f(x_1, x_2, \dots, x_n, c)$ in the case of the Naive Bayes model. Thus, from the earlier equation, we can state that $f_{\mathcal{G}_{NB}} = f_{\mathcal{G}_T}$ under the zero-one loss function when $f_{\mathcal{G}_T} \geq 1$ and $\prod_{i=1}^{n} ddr_{\mathcal{G}_T}(x_i) \leq f_{\mathcal{G}_T}$, or $f_{\mathcal{G}_T} < 1$ and $\prod_{i=1}^{n} ddr_{\mathcal{G}_T}(x_i) > f_{\mathcal{G}_T}$. Therefore, we can conclude that the distribution of dependencies between the features over classes (that is $ddr_{\mathcal{G}_T}(x_i)$) affect the classification, not merely the dependencies.

So, in cases where for each feature, x_i, $ddr_{g_T}(x_i) = 1$, that is, the local distribution of each feature is distributed evenly across both positive and negative classes, the Naive Bayes model will perform in the same way as the model representing the dependencies among the features. Further, in cases where $\prod_{i=1}^{n} ddr_{g_T}(x_i) = 1$, that is, the influence of some local dependencies in favor of class c^+ is canceled by the influence of some other dependencies in favor of class c^-, Naive Bayes will be an optimal classifier as well.

Because of its independence assumption, the parameters for each feature can be learned separately, which greatly simplifies the learning process and is very useful in a domain where we have very many features. In the case of document classification, the features or the attributes of a document are nothing but the words comprising it. In most of the cases, the vocabulary is huge, thus leading to a very large number of features. So, one of the major algorithms used in document classification is Naive Bayes.

Types of Naive Bayes models

There are two variants of the Naive Bayes model that are generally used for document classification.

One model specifies that a document is represented by a vector of binary attributes indicating which words occur and which words do not occur in the document. The attributes are independent of the number of times a word occurs in a document. So, the computation of the probability of a document involves multiplication of the probabilities of all the attribute values, including the probability of non-occurrence for words that do not occur in the document. Here, we consider the document to be the event, and the absence or presence of words to be attributes of the event. This describes a distribution based on a multivariate Bernoulli event model.

The second model specifies that a document is represented by the set of word occurrences in the document. In this model, however, the number of occurrences of each word in the document is captured. So, computing the probability of a document involves multiplication of the probability of the words that occur. Here we consider the individual word occurrences to be the events and the document to be a collection of word events. We call this a multinomial event model.

Multivariate Bernoulli Naive Bayes model

As stated earlier, in the case of a multivariate Bernoulli Naive Bayes model, a document is considered as a binary vector of the space of words for a given vocabulary V. The document d can be represented as $\{b_1, b_2, \ldots, b_{|V|}\}$, where b_i corresponds to the presence of the word w_i in the document; $b_i = 1$ if $b_i = 0$ is present, b_{it} otherwise. More often, w_t is defined as an indicator variable representing the presence of the word d_i in the document .

Thus, $P(d_i | c_j)$ is defined as follows:

$$P(d_i | c_j) = \prod_{t=1}^{|V|} \left(b_{it} \cdot P(w_t | c_j) \right) + \left((1 + b_{it}) \cdot (1 - P(w_t | c_j)) \right)$$

We can see that a document is considered as a collection of multiple independent Bernoulli experiments, one for each word in the vocabulary, with the probabilities for each of these word events defined by each component $P(w_t | c_j)$.

The `scikit-learn` Python module provides us with the implementation of the Naive Bayes model. Let's look at an example of text classification. Before going into the details of classification, let's discuss one of the major steps in text classification known as feature extraction.

The most common strategy to extract features from a text document is called a bag-of-words representation. In this representation, documents are described by word occurrences while completely ignoring the relative position information of the words in the document. The `scikit-learn` Python module provides utilities for the most common ways of extracting numerical features from text content, which includes the following:

- Tokenizing strings and giving an integer ID for each possible token
- Counting the occurrences of tokens in each document
- Normalizing and weighting with diminishing importance of tokens that occur in the majority of samples/documents

Let's discuss the various feature extraction methods implemented in `scikit-learn` with examples. The first feature extractor we will discuss is `CountVectorizer`. It implements both tokenization and counting occurrences:

```
In [1]: from sklearn.feature_extraction.text import CountVectorizer
# The input parameter min_df is a threshold which is used to
# ignore the terms that document frequency less than the
# threshold. By default it is set as 1.
```

```
In [2]: vectorizer = CountVectorizer(min_df=1)

In [3]: corpus = ['This is the first document.',
                  'This is the second second document.',
                  'And the third one.',
                  'Is this the first document?']

# fit_transform method basically Learn the vocabulary dictionary
# and return term-document matrix.
In [4]: X = vectorizer.fit_transform(corpus)

# Each term found by the analyzer during the fit is assigned a
# unique integer index corresponding to a column in the resulting
# matrix.
In [5]: print(vectorizer.get_feature_names())
   ['and', 'document', 'first', 'is', 'one', 'second', 'the',
    'third', 'this'])

# The numerical features can be extracted by the method toarray
# It returns a matrix in the form of (n_corpus, n_features)
# The columns correspond to vectorizer.get_feature_names(). The
# value of a[i, j] is basically the count of word correspond to
# column j in document i.
In [6]: print(X.toarray())
array([[0, 1, 1, 1, 0, 0, 1, 0, 1],
       [0, 1, 0, 1, 0, 2, 1, 0, 1],
       [1, 0, 0, 0, 1, 0, 1, 1, 0],
       [0, 1, 1, 1, 0, 0, 1, 0, 1]]...)

# Instead of using the count we can also get the binary value
# matrix for the given corpus by setting the binary parameter
# equals True.
In [7]: vectorizer_binary = CountVectorizer(min_df=1, binary=True)

In [8]: X_binary = vectorizer_binary.fit_transform(corpus)
# The value of a[i, j] == 1 means that the word corresponding to
# column j is present in document i
In [9]: print(X_binary.toarray())
   array([[0, 1, 1, 1, 0, 0, 1, 0, 1],
          [0, 1, 0, 1, 0, 1, 1, 0, 1],
          [1, 0, 0, 0, 1, 0, 1, 1, 0],
          [0, 1, 1, 1, 0, 0, 1, 0, 1]])
```

Another interesting feature extractor is called *tf-idf*, which is short for term "frequency-inverse document frequency". In a large text corpus, some words are present in almost all the documents. These include "the", "a", and "is", hence carrying very little meaningful information about the actual contents of the document. If we were to feed the direct count data directly to a classifier, these very frequent terms would shadow the frequencies of rarer yet more interesting terms. Tf-idf basically diminishes the importance of these words that occur in the majority of documents. It is basically the product of two terms, namely term frequency and inverse document frequency. Term frequency corresponds to the frequency of a term in a document, that is, the number of times the term *t* appears in the document *d*. Inverse document frequency is a measure of how much information the word provides, that is, whether the term is common or rare across all documents. Let's take an example for Tf-idf using scikit-learn:

```
In [1]: from sklearn.feature_extraction.text import TfidfVectorizer

In [2]: corpus = ['This is the first document.',
                  'This is the second second document.',
                  'And the third one.',
                  'Is this the first document?']

# The input parameter min_df is a threshold which is used to
# ignore the terms that document frequency less than the
# threshold. By default it is set as 1.
In [3]: vectorizer = TfidfVectorizer(min_df=1)

# fit_transform method basically Learn the vocabulary dictionary
# and return term-document matrix.
In [4]: X = vectorizer.fit_transform(corpus)

# Each term found by the analyzer during the fit is assigned a
# unique integer index corresponding to a column in the resulting
# matrix.
In [5]: print(vectorizer.get_feature_names())
['and', 'document', 'first', 'is', 'one', 'second', 'the',
 'third', 'this'])

# The numerical features can be extracted by the method toarray
# It returns a matrix in the form of (n_corpus, n_features)
# The columns correspond to vectorizer.get_feature_names(). The
# value of a[i, j] is basically the count of word correspond to
# column j in document i
In [6]: print(X.toarray())
[[ 0.          0.43877674  0.54197657  0.43877674  0.          0.
   0.35872874  0.          0.43877674]
 [ 0.          0.27230147  0.          0.27230147  0.          0.85322574
```

```
      0.22262429  0.            0.27230147]
 [ 0.55280532  0.            0.            0.            0.55280532  0.
   0.28847675  0.55280532  0.            ]
 [ 0.           0.43877674  0.54197657  0.43877674  0.            0.
   0.35872874  0.            0.43877674]]
```

There are other implementations of feature extractors in scikit-learn, such as HashingVectorizer, which uses the hashing trick to create a mapping from the string token name to the feature index. It turns a collection of text documents into a scipy.sparse matrix holding token occurrence counts. As it uses the scipy.sparse matrix, it is very memory efficient and can be used in the case of large text documents.

Let's come back to our discussion on the implementation of text classification using the multivariate Bernoulli Naive Bayes model:

```
In [1]: from sklearn.datasets import fetch_20newsgroups
In [2]: from sklearn.feature_extraction.text import HashingVectorizer
In [3]: from sklearn.feature_extraction.text import CountVectorizer
In [4]: from sklearn.naive_bayes import BernoulliNB
In [5]: from sklearn import metrics


# The dataset used in this example is the 20 newsgroups dataset.
# The 20 Newsgroups data set is a collection of
# approximately 20,000 newsgroup documents, partitioned (nearly)
# evenly across 20 different newsgroups. It will be
# automatically downloaded, then cached.


# For our simple example we are only going to use 4 news group
In [6]: categories = ['alt.atheism',
                      'talk.religion.misc',
                      'comp.graphics',
                      'sci.space']


# Loading training data
In [7]: data_train = fetch_20newsgroups(subset='train',
                                        categories=categories,
                                        shuffle=True,
                                        random_state=42)


# Loading test data
In [8]: data_test = fetch_20newsgroups(subset='test',
                                       categories=categories,
                                       shuffle=True,
                                       random_state=42)
```

```
In [9]: y_train, y_test = data_train.target, data_test.target

# It can be changed to "count" if we want to use count vectorizer
In [10]: feature_extractor_type = "hashed"

In [11]: if feature_extractor_type == "hashed":
             # To convert the text documents into numerical features,
             # we need to use a feature extractor. In this example we
             # are using HashingVectorizer as it would be memory
             # efficient in case of large datasets
                 vectorizer = HashingVectorizer(stop_words='english')

                 # In case of HashingVectorizer we don't need to fit
                 # the data, just transform would work.
                 X_train = vectorizer.transform(data_train.data)
                 X_test = vectorizer.transform(data_test.data)

             elif feature_extractor_type == "count":
             # The other vectorizer we can use is CountVectorizer with
             # binary=True. But for CountVectorizer we need to fit
             # transform over both training and test data as it
             # requires the complete vocabulary to create the matrix
                 vectorizer = CountVectorizer(stop_words='english',
                                              binary=True)

     # First fit the data
In [12]: vectorizer.fit(data_train.data + data_test.data)

# Then transform it
In [13]: X_train = vectorizer.transform(data_train.data)
In [14]: X_test = vectorizer.transform(data_test.data)

# alpha is additive (Laplace/Lidstone) smoothing parameter (0 for
# no smoothing).
In [15]: clf = BernoulliNB(alpha=.01)

# Training the classifier
In [16]: clf.fit(X_train, y_train)

# Predicting results
In [17]: y_predicted = clf.predict(X_test)

In [18]: score = metrics.accuracy_score(y_test, y_predicted)
In [19]: print("accuracy: %0.3f" % score)
```

Multinomial Naive Bayes model

In the previous section, we discussed the multivariate Bernoulli Naive Bayes model. In this section, we are going to discuss another variant called the multinomial model. Unlike the previous model, it captures the word frequency information of a document.

In the multinomial model, a document is considered to be an ordered sequence of word events drawn from the same vocabulary V. Again, we make a similar Naive Bayes assumption that the probability of each word event in a document is independent of the word's context and position in the document. Thus, each document d_i is drawn from a multinomial distribution of words with as many independent trials as the length of d_i. The distribution is parameterized by vectors $\theta_c = \left(\theta_{c1}, \theta_{c2}, \ldots, \theta_{c|V|}\right)$ for all $c \in C$, where $|V|$ is the size of the vocabulary and θ_{ci} represents the probability of the word w_i belonging to the class c, that is $P\left(w_i \,|\, c\right)$.

The parameter θ_{ci} is estimated by a maximum likelihood estimate as follows:

$$\theta_{ci} = \frac{N_{ci} + \alpha}{N_c + |V|\alpha}$$

Here, N_{ci} is defined as the number of times the word w_i appeared in the sample of class c in the training set $N_{ci} = \sum_{d \in T[c]} x_{di}$, where x_{di} represents the word count of w_i in the document d, $T[c]$ represents all the samples of the training set T belonging to the class c, and N_c is defined as the total count of all features for class c, that is $N_c = \sum_{i=1}^{|V|} N_{ci}$.

The smoothing parameter α accounts for features not present in the learning samples. It prevents the assignment of zero probabilities to words not present in a particular class. Setting $\alpha = 1$ is called Laplace smoothing, while $\alpha < 1$ is called Lidstone smoothing.

It's implementation in Python is as follows:

```
In [1]: from sklearn.datasets import fetch_20newsgroups
In [2]: from sklearn.feature_extraction.text import TfidfVectorizer
In [3]: from sklearn.feature_extraction.text import CountVectorizer
In [4]: from sklearn.naive_bayes import MultinomialNB
In [5]: from sklearn import metrics

# Just like the previous example, here also we are going to deal
# 20 newsgroup data.
```

```
In [6]: categories = ['alt.atheism',
                       'talk.religion.misc',
                       'comp.graphics',
                       'sci.space']

# Loading training data
In [7]: data_train = fetch_20newsgroups(subset='train',
                                         categories=categories,
                                         shuffle=True,
                                         random_state=42)

# Loading test data
In [8]: data_test = fetch_20newsgroups(subset='test',
                                        categories=categories,
                                        shuffle=True,
                                        random_state=42)

In [9]: y_train, y_test = data_train.target, data_test.target
In [10]: feature_extractor_type = "tfidf"

In [11]: if feature_extractor_type == "count":
            # The other vectorizer we can use is CountVectorizer
            # But for CountVectorizer we need to fit transform over
            # both training and test data as it requires the complete
            # vocabulary to create the matrix
                vectorizer = CountVectorizer(stop_words='english')
                vectorizer.fit(data_train.data + data_test.data)
                X_train = vectorizer.transform(data_train.data)
                X_test = vectorizer.transform(data_test.data)

            elif feature_extractor_type == "tfidf":
                vectorizer = TfidfVectorizer(stop_words="english")
                X_train = vectorizer.fit_transform(data_train.data)
                X_test = vectorizer.transform(data_test.data)

# alpha is additive (Laplace/Lidstone) smoothing parameter (0 for
# no smoothing).
In [12]: clf = MultinomialNB(alpha=.01)

# Training the classifier
In [13]: clf.fit(X_train, y_train)

# Predicting results
In [14]: y_predicted = clf.predict(X_test)

In [15]: score = metrics.accuracy_score(y_test, y_predicted)
In [16]: print("accuracy: %0.3f" % score)
```

Choosing the right model

In the previous sections, we discussed two different variants of Naive Bayes models, the multivariate Bernoulli model and the multinomial model. There has been a lot of research on which model to choose. McCallum and Nigam (1998) did extensive comparisons of both the models (refer to the research paper titled *A Comparison of Event Models for Naive Bayes Text Classification*). They found that the multivariate Bernoulli model performs well with small vocabulary sizes, but the multinomial Bernoulli model usually performs better at larger vocabulary sizes, providing on average, a 27 percent reduction in error over the multivariate Bernoulli model at any vocabulary size. However, it is advisable to evaluate both the models.

Dynamic Bayesian networks

In the examples we have seen so far, we have mainly focused on *variable-based models*. In these types of models, we mainly focus on representing the variables of the model. As in the case of our restaurant example, we can use the same network structure for multiple restaurants as they share the same variables. The only difference in all these networks would be the different states in the case of different restaurants. These types of models are known as variable-based models.

Let's take a more complex example. Let's say we want to model the state of a robot traveling over some trajectory. In this case, the state of the variables will change with time, and also, the states of some variables at some instance t might depend on the state of the robot at instance $t-1$. Clearly, we can't model such a situation with a variable-based model. So, generally, for such problems, we use **dynamic Bayesian networks (DBNs)**.

Assumptions

Before discussing the simplifying assumptions that DBNs make, let's first see the notations that we are going to use in the case of DBNs. As DBNs are defined over a range of time, with each time instance having the same variables, representing the instantiation of a random variable X_i at a time instance t, we will be using $X_i^{(t)}$. The variable X_i is now known as a template variable as it can't take any values itself. This template variable is instantiated at various time instances, and at each instance t, the variable $X_i^{(t)}$ can take values from $Val(X_i)$. Also, for a set of random variables $\mathbf{X} \subseteq \chi$, we use $\mathbf{X}^{(t_1:t_2)}$, where $t_1 < t_2$ to denote the set of variables $\{\mathbf{X}^{(t)} : t \in [t_1, t_2]\}$. Similarly, we use the notation to denote the assignments to this set of variables.

As we can see, the number of variables will be huge between any considerable time difference and hence, our joint distribution over such trajectories will be very complex. Therefore, we make some assumptions to simplify our distribution.

Discrete timeline assumption

The first simplifying assumption that we make is to have a discrete timeline rather than having a continuous one. So, the measurement of the states of the random variables are taken at some predetermined time interval Δ. With this assumption now, the random variable $\chi^{(t)}$ represents the values of the variables at a time instance $t \cdot \Delta$.

Using this assumption, we can now write the distribution over the variable over a time period 0 to T as follows:

$$P\left(\chi^{(0:T)}\right) = \prod_{t=0}^{T-1} P\left(\chi^{(t+1)} \mid \chi^{(0:t)}\right)$$

Therefore, the distribution over trajectories is the product of conditional distribution over the variables at each previous time instance, given all the past variables.

The Markov assumption

The second assumption that we make is as follows:

$$\chi^{(t+1)} \perp \chi^{(0:(t-1))} \mid \chi^{(t)}$$

Putting this in simple words, the variables at time $t + 1$ can directly depend only on the variables at time t and are thus, independent of all the variables $\chi^{(t')}$ for $t' < t - 1$. Any system that satisfies this condition is known as *Markovian*. This assumption reduces the earlier joint distribution equation to the following:

$$P\left(\chi^{(0:T)}\right) = \prod_{t=0}^{T-1} P\left(\chi^{(t+1)} \mid \chi^{(t)}\right)$$

In other words, this assumption also constraints our network, such that the variables in $\chi^{(t+1)}$ can't have any edges from any other variable in $\chi^{(0:t-1)}$.

However, the problem with this assumption is that it may not hold in all cases. Let's take an example to show this. Suppose we want to model the location of a car. As we can see, we can easily predict the location of the car in the future, given the observations about the past. Also, let's assume that we only have two random variables {L,O} and L representing the location of the car and O representing the observed location. Here, we might think that our model satisfies the Markov assumption as the location at *t + 1* will only depend on the location at time *t* and is independent of the location at t' for $t' < t$. However, this intuition might turn out to be wrong as we don't know the velocity or the direction of travel of the car. Had we known the previous locations of the car, we could have easily estimated both the direction and velocity. So, in such cases, to make our model closer to satisfying our Markov assumption, we can add the variables `direction` and `velocity` in our model. Now, at each instance of time, if we know the velocity and direction of motion of the car, we can predict the next instance using just the values of the previous instance. Now, to account for the changes in the velocity and direction, we can also add variables such as weather conditions and road conditions. With the addition of these extra variables, our model is now close to being Markovian.

Model representation

The Markov assumption and the independence assumption that we saw in the previous section allow us to represent the joint distribution very compactly, even over infinite trajectories. All we need to define is the distribution for the initial state and a transition model $P(\chi' | \chi)$. We can represent the preceding car example using a network as shown in Fig 7.4, Fig 7.5, and Fig 7.6.

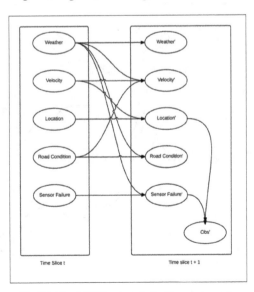

Fig 7.4: The 2-TBN network for the car example

The following flowchart depicts the network structure at time $t = 0$:

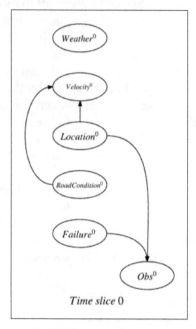

Fig 7.5: The network structure

The following figure is the flowchart that shows the unrolled DBN over a two-time slice:

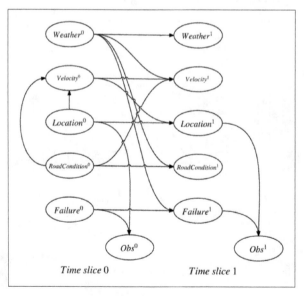

Fig 7.6: Unrolled DBN over a two-time slice

Also, we define the interface variables χI as variables whose values at time t have a direct effect on the variables at time $t + 1$. Therefore, only the variables in χI can be parents of the variables in χ'. Also, the preceding car example is an example of a two-time slice **Bayesian network (2-TBN)**. We define a 2-TBN for a process over χ as a conditional Bayesian network over χ', given χI, where $\chi I \subseteq \chi$ is a set of interface variables. In our example, all the variables are interface variables, except for O.

Overall, this 2-TBN represents the following conditional distribution:

$$P(\chi' \mid \chi) = P(\chi' \mid \chi I) = \prod_{i=1}^{n} P\left(X_i' \mid Pa_{X_i'}\right)$$

For each template variable X_i, the CPD $P\left(X_i' \mid Pa_{X_i'}\right)$ is known as the template factor. This template factor is instantiated multiple times in the network for each $X_i^{(t)}$.

Currently, none of the Python libraries for PGM has a concrete implementation to work with DBN. However, pgmpy developers are currently working on it so it should soon be available in pgmpy.

The Hidden Markov model

In the previous section, we discussed DBNs. In this section, we will discuss one particular variant of it, called the **Hidden Markov model (HMM)**. Although named the Hidden Markov model, it is not a Markov network. Its etymology comes from the fact that the HMM satisfies the Markov property.

A Markov property basically indicates the memory-less property of a stochastic process, and any stochastic process satisfying this property is called as a Markov process. Let $\{X(t), t \geq 0\}$ be a time-continuous process. Then, for every $n \geq 0$, time points $0 \leq t_0 < t_1 < \cdots < t_{n-1} < t_n$ with states $i_0, i_1, \cdots i_n$. Then, $P\left(X(t_n) = i_n \mid X(t_n - 1) = i_{n-1}, \cdots, X(t_0) = i_0\right) = P\left(X(t_n) = i_n \mid X(t_n - 1) = i_{n-1}\right)$. This means that the current state depends only on the previous state; any additional knowledge about the history doesn't add any extra information.

For example, if we sample the mood of a person once a minute, then it is fair to assume that the current mood of the person is only affected by his/her mood in the previous minute (unless that person is suffering from bipolar disorder). In the case of predicting the trajectory of a missile, we can also assume that the position of the missile at X_{t+1} can be determined by X_t alone. Although at first glance, this may not seem to be correct, if the trajectory is sampled very fast, it may be a very good approximation.

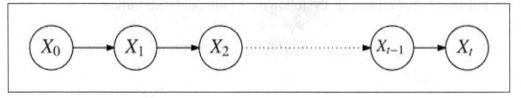

Fig 7.7: Graphical model representation of a Markov process

Fig 7.7 shows the graphical model representation of a Markov process. In most applications of such models, the probability distribution $P(X_t \mid X_{t-1})$ is assumed to be equal for any value of t. $P(X_t \mid X_{t-1})$ can be represented in the form of a transition matrix (A) or a state-transition diagram. For example, if we want to model the mood of a person (which can be very sad, sad, happy, or very happy), we can represent this in the form of a state-transition diagram:

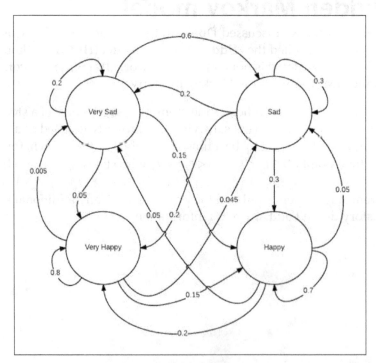

Fig 7.8: State-transition diagram representing the transition of the mood of a person across time

The preceding figure shows the transition of the mood of a person from one state to another. For example, from the diagram, we can infer that the probability of transitioning from a very happy state to a happy state is 0.15, the probability of remaining in the same state is 0.8, and so on. As all the edge weightings represent the probability, all the weightings corresponding to the edges outgoing from a single node should sum up to 1.

Another way of representing $P(X_t \mid X_{t-1})$ is with a transition matrix. A transition matrix (A) is a matrix in the shape of $N \times N$, where N represents the number of states. Each element a_{ij} of a transition matrix A represents the probability of transitioning from state s_i to state s_j. For example, the transition matrix corresponding to the preceding state-transition diagram would be as follows:

	Very Sad	Sad	Happy	Very Happy
Very Sad	0.2	0.6	0.15	0.05
Sad	0.2	0.3	0.3	0.2
Happy	0.05	0.05	0.7	0.2
Very Happy	0.005	0.045	0.15	0.8

For the first node X_0, there is no parent node. So unlike all other nodes, its distribution can't be encoded by the conditional probability distribution of the form $P(X_t \mid X_{t-1})$. So, for this node, the distribution is a marginal probability distribution called the initial state probability distribution π. It is an array of shape of $N \times 1$, with the constraint $\sum_{i=1}^{N} \pi_i = 1$. For example, in the earlier mood example, the matrix can be as follows:

Very Sad	Sad	Happy	Very Happy
0.1	0.4	0.4	0.1

However, in real-life situations, we can't directly observe the state of the variable, that is, the variables are hidden from us. For example, we can't observe whether a person is very sad, sad, happy, or very happy just by looking at this table. Instead, we can observe some other variable $X_1, X_2, \cdots, X_t$ that is affected by Z_t. For example, the current activity of a person (which is an observable parameter) can tell us about his/her mood. Thus, the graphical model representing the system, as stated in Fig 7.7, is modified as follows:

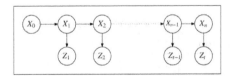

Fig 7.9: Graphical model representing a Hidden Markov model

With the addition of extra nodes and edges to the graphical model, we need an additional conditional probability distribution $P(Z_t \mid X_t)$ (called emission probability), which is represented as Θ. It is assumed to be equal for any value of t. Thus, an HMM model can be represented by the following three parameters:

- The initial state probability distribution $(P(X_0))$, represented as $\lambda = (A, \Theta, \pi)$
- The transition matrix corresponds to the distribution $P(X_t \mid X_{t-1})$ and is represented as A
- Emission probabilities corresponding to the distribution and are represented as Θ

Thus, an HMM model can be stated as $\lambda = (A, \Theta, \pi)$.

Generating an observation sequence

Given model $\lambda = (A, \Theta, \pi)$, we can generate a sequence of observations $\mathbf{Z} = \{Z_1, Z_2, \cdots, Z_T\}$ as follows:

1. Choose an initial state $X_0 = s_i$ ($s_i \in \{1, \cdots, N\}$) according to the initial state distribution π.
2. Set $t = 1$.
3. Choose an observation O_t corresponding to X_t according to the emission probability Θ.
4. Transit to the next state X_{t+1} according to the state-transition probability represented by the transition matrix A.
5. Set $t = t + 1$ and return to step 3 if $t < T$, else terminate.

For HMM and its application in Python, we will use a library called `hmmlearn`. It is an offshoot of a popular machine learning library in Python called `scikit-learn`.

Let's continue with the previous mood example. Suppose we are able to observe the current activity of a person, and for the sake of simplicity, let's assume it is restricted to a few possibilities such as watching television, sleeping, eating, crying, and playing. As the observed value is also a discrete quantity, the emission probability Θ can be represented in the form of a tabular conditional probability distribution.

	Very Sad	Sad	Happy	Very Happy
Watching Television	0.045	0.2	0.3	0.1
Sleeping	0.15	0.2	0.1	0.1

	Very Sad	Sad	Happy	Very Happy
Eating	0.2	0.2	0.1	0.2
Crying	0.6	0.3	0.05	0.05
Playing	0.005	0.1	0.45	0.55

The preceding distribution represents the probability given the mood of the person. For example, the first row and first column basically represent the probability of someone watching television when he/she is very sad.

To represent an HMM with multinomial (or discrete) emission, The `hmmlearn` library provides a class called `MultinomialHMM`. It's implementation in Python is as follows:

```
In [1]: from hmmlearn.hmm import MultinomialHMM
In [2]: import numpy as np

# Here n_components correspond to number of states in the hidden
# variables and n_symbols correspond to number of states in the
# obversed variables
In [3]: model_multinomial = MultinomialHMM(n_components=4)

# Transition probability as specified above
In [4]: transition_matrix = np.array([[0.2, 0.6, 0.15, 0.05],
                                      [0.2, 0.3, 0.3, 0.2],
                                      [0.05, 0.05, 0.7, 0.2],
                                      [0.005, 0.045, 0.15, 0.8]])
# Setting the transition probability
In [5]: model_multinomial.transmat_ = transition_matrix

# Initial state probability
In [6]: initial_state_prob = np.array([0.1, 0.4, 0.4, 0.1])

# Setting initial state probability
In [7]: model_multinomial.startprob_ = initial_state_prob

# Here the emission prob is required to be in the shape of
# (n_components, n_symbols). So instead of directly feeding the
# CPD we would using the transpose of it.
In [8]: emission_prob = np.array([[0.045, 0.15, 0.2, 0.6, 0.005],
                                  [0.2, 0.2, 0.2, 0.3, 0.1],
                                  [0.3, 0.1, 0.1, 0.05, 0.45],
                                  [0.1, 0.1, 0.2, 0.05, 0.55]])

# Setting the emission probability
```

```
In [9]: model_multinomial.emissionprob_ = emission_prob

# model.sample returns both observations as well as hidden states
# the first return argument being the observation and the second
# being the hidden states
In [10]: Z, X = model_multinomial.sample(100)
```

The other type of HMM model that implements in `hmmlearn` is `GaussianHMM`. It represents HMM with Gaussian emissions. Thus, for characterizing the emission probability Θ, instead of using a complete tabular CPD, we can just provide the mean and covariance. For example, let's try to sample observations from an HMM with $N = 3$ and with a mean μ and covariance Σ:

```
In [1]: from hmmlearn.hmm import GaussianHMM
In [2]: import matplotlib.pyplot as plt
In [3]: import numpy as np

# Here n_components correspond to number of states in the hidden
# variables.
In [4]: model_gaussian = GaussianHMM(n_components=3,
                                     covariance_type='full')

# Transition probability as specified above
In [5]: transition_matrix = np.array([[0.2, 0.6, 0.2],
                                      [0.4, 0.3, 0.3],
                                      [0.05, 0.05, 0.9]])

# Setting the transition probability
In [6]: model_gaussian.transmat_ = transition_matrix

# Initial state probability
In [7]: initial_state_prob = np.array([0.1, 0.4, 0.5])

# Setting initial state probability
In [8]: model_gaussian.startprob_ = initial_state_prob

# As we want to have a 2-D gaussian distribution the mean has to
# be in the shape of (n_components, 2)
In [9]: mean = np.array([[0.0, 0.0],
                         [0.0, 10.0],
                         [10.0, 0.0]])

# Setting the mean
In [10]: model_gaussian.means_ = mean
```

```
# As emission probability is a 2-D gaussian distribution, thus
# covariance matrix for each state would be a 2-D matrix, thus
# overall the covariance matrix for all the states would be in the #
form of (n_components, 2, 2)
In [11]: covariance = 0.5 * np.tile(np.identity(2), (3, 1, 1))
In [12]: model_gaussian.covars_ = covariance

# model.sample returns both observations as well as hidden states
# the first return argument being the observation and the second
# being the hidden states
In [13]: Z, X = model_gaussian.sample(100)

# Plotting the observations
In [14]: plt.plot(Z[:, 0], Z[:, 1], "-o", label="observations",
                ms=6, mfc="orange", alpha=0.7)

# Indicate the state numbers
In [15]: for i, m in enumerate(mean):
            plt.text(m[0], m[1], 'Component %i' % (i + 1),
                    size=17, horizontalalignment='center',
                    bbox=dict(alpha=.7, facecolor='w'))

In [16]: plt.legend(loc='best')
In [17]: plt.show()
```

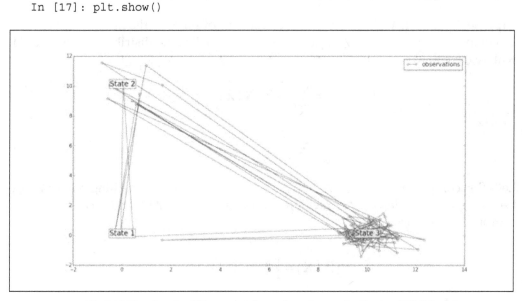

Fig 7.10: Plot showing 100 samples drawn from the previously stated HMM.
The lines connect the successive observations.

Fig 7.10 shows the successive observations drawn from the HMM stated earlier. In this HMM, the initial state probability distribution favors the state s_3 as compared to the other two states. According to the transition matrix, the probability of transitioning state s_3 to s_3 is much higher as compared to transitioning from s_3 to any other state. Thus, we can see that most of the observations correspond to the state s_3 as compared to any other state.

Computing the probability of an observation

The next problem that we are going to tackle in the case of the HMM is computing the probability of observation given a model that is computing $P(\mathbf{Z}, \mathbf{X} \mid \lambda)$.

Let's start with a simple example of an HMM with a multinomial emission (that is, the observation variable being discrete quantities). In this case, the emission probability Θ can be represented by a matrix B such that each element b_{ij} equals the following:

$$b_{ij} = P(Z_t = j \mid X_t = i) \quad 1 \leq i \leq N$$
$$1 \leq j \leq M$$

Here, N represents the number of possible states of a hidden variable and M represents the number of possible states of an observed variable.

Suppose $\mathbf{X} = Z\{X_0, X_1, X_2, \cdots, X_T\}$ is the sequence of states of the hidden variable. To compute the value of $P(\mathbf{Z}, \mathbf{X} \mid \lambda)$, we can marginalize the distribution $P(\mathbf{Z}, \mathbf{X} \mid \lambda)$ with respect to X:

$$P(\mathbf{Z} \mid \lambda) = \sum_{\mathbf{X}} P(\mathbf{Z}, \mathbf{X} \mid \lambda)$$
$$= \sum_{\mathbf{X}} \left(P(\mathbf{Z} \mid \mathbf{X}, \lambda) \cdot P(\mathbf{X} \mid \lambda) \right)$$

Let's first compute the term $P(\mathbf{Z} \mid \mathbf{X}, \lambda)$. The $P(\mathbf{Z} \mid \mathbf{X}, \lambda)$ term is nothing but $P(\mathbf{Z} \mid \mathbf{X})$, because given a model λ, Z_t only depends on X_t. The $P(\mathbf{Z} \mid \mathbf{X})$ term can be computed as follows:

$$P(\mathbf{Z} \mid \mathbf{X}) = \prod_{i=1}^{T} P(Z_i \mid X_i)$$
$$= b_{X_1 Z_1} \cdot b_{X_2 Z_2} \cdots b_{X_T Z_T}$$

The $P(\mathbf{X}|\lambda)$ term can be computed directly from the initial state probability distribution and transition matrix as follows:

$$P(\mathbf{X}|\lambda) = P(X_0)\prod_{i=1}^{T}P(X_i|X_{i-1})$$

$$= \pi_{X_0} \cdot a_{X_0 X_1} \cdot a_{X_2 X_1} \cdots a_{X_{T-1} X_T}$$

Thus, $P(\mathbf{Z}|\lambda)$ can be stated in the following way:

$$P(\mathbf{Z}|\lambda) = \sum_{\mathbf{X}}\left(P(\mathbf{Z}|\mathbf{X},\lambda)\cdot P(\mathbf{X}|\lambda)\right)$$

$$= \sum_{X_1,X_2,\cdots,X_T} \pi_{X_0} \cdot a_{X_0 X_1} \cdot b_{X_1,Z_1} \cdots a_{X_{T-1}} b_{X_T Z_T}$$

The computation of $P(\mathbf{Z}|\lambda)$ using the preceding equation requires an exponentially large number of mathematical operations, precisely $(2T-1)\cdot N^T$ multiplications and $N^T - 1$ additions. Even for a very small value of T (for example, 100) and N as 5, it requires $2 \cdot 100 \cdot 5^{1}00 \approx 10^{72}$ operations. Thus, we require a more efficient way to compute $P(\mathbf{Z}|\lambda)$. One such method is the forward-backward algorithm.

The forward-backward algorithm

Before going into the details of the algorithm, let's define some variables that are needed for this routine, the first one being the forward variable $\alpha_i(t)$. It is defined as follows:

$$\alpha_i(t) = P(Z_1,Z_2,\cdots,Z_t,X_t = s_i | \lambda)$$

The forward variable is the probability of a partially observed sequence $\{Z_1,Z_2,\cdots,Z_t\}$ (until a time t) and the state s_i at time t, given the model λ. $\alpha_i(t)$, can be computed inductively as the following initialization:

$$\alpha_1(t) = \pi_i \cdot b_i z_1 \forall i \in [1,N]$$

Here, π_i represents the initial probability for the state s_i and $b_i z_1$ represents the probability of Z_1 given the state of $X_1 = s_i$.

The induction step:

$$\alpha_{t+1}(j) = \left[\sum_{i=1}^{N} \alpha_t(i) a_{ij} \right] b_j z_{t+1}, \quad 1 \le t < T$$

$$1 \le j \le N$$

Here, a_{ij} represents the probability of transitioning from the state s_i to the state s_j and $b_j z_{t+1}$ represents the probability of Z_{t+1} given $X_1 = s_j$.

The termination step:

$$P(\mathbf{Z} \mid \lambda) = \sum_{i=1}^{N} P(Z_1, Z_2, \cdots, Z_T, X_T = s_i \mid \lambda)$$

$$= \sum_{i=1}^{N} \alpha_T(i)$$

The induction step is the core of the computational method.

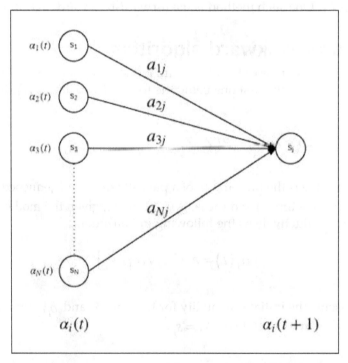

Fig 7.11: Computation of $\alpha_{t+1}(j)$ as shown in the induction step

Fig 7.11 shows the computation algorithm used in the induction step. The values of all the $\alpha_i(t)$ instances are weight-summed, where the weightings represent the probability of transitioning from the state s_i to the state s_j.

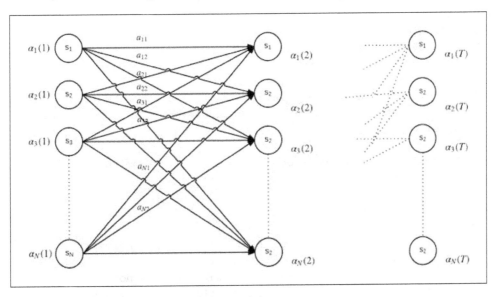

Fig 7.12: Implementation of the computation of $\alpha_{t+1}(j)$ in terms of the lattice of observation t and states i

This operation only requires $N^2 \cdot T$ operations, as opposed to the $2T \cdot N^T$ operations required by the direct calculation. So, in the case of $N = 5$ and $T = 100$, we only require 3000 computations.

In a similar manner, we can use a backward pass to compute the backward variable $\beta_t(i)$. This is defined as follows:

$$\beta_t(i) = P(Z_{t+1}, Z_{t+2}, \cdots, Z_T \mid X_t = s_i, \lambda)$$

The initialization step:

$$\beta_i(T) = 1 \forall i \in [1, N]$$

The induction step:

$$\beta_i(t) = \sum_{j=1}^{N} a_{ij} b_j z_{t+1} \beta_{t+1}(j), \quad t = T-1, T-2, \cdots, 1$$

$$1 \le j \le N$$

The `hmmlearn` module facilitates the computation of $P(\mathbf{Z}|\lambda)$. For example, let's take the previously stated example of the `GaussianHMM` model:

```
# mean of the emission probability distribution for various states
# were:
# [0.0, 0.0],
# [0.0, 10.0],
# [10.0, 0.0]


# So if an observations are sampled from some other gaussian
# distribution with mean centered at different location such as:
# [5, 5]
# [-5, 0]
# the probability of these observations coming from this model
# should be very low.


# generating observations
In [18]: observations = np.row_stack((
                        np.random.multivariate_normal(
                            [5, 5], [[0.5, 0], [0, 0.5]], 10),
                        np.random.multivariate_normal(
                            [-5, 0], [[0.5, 0], [0, 0.5]], 10)))


# model.score returns the log-probability of P(observations |
# model)
In [19]: score_1 = model_gaussian.score(observations)
In [20]: print(score_1)
-728.50717880180241


# Lets try to check whether observations sampled from the
# multivariate normal distributions that were used in our HMM
# model provides greater value of score or not
In [21]: observations = np.row_stack((
                        np.random.multivariate_normal(
                            [10, 0], [[0.5, 0], [0, 0.5]], 10),
                        np.random.multivariate_normal(
                            [0, 0], [[0.5, 0], [0, 0.5]], 2),
                        np.random.multivariate_normal(
                            [0, 10], [[0.5, 0], [0, 0.5]], 4)))
In [22]: score_2 = model_gaussian.score(observations)
In [23]: print(score_2)
-44.709532774805481


# We can see that results matches our intuition
```

Computing the state sequence

Apart from computing $P(\mathbf{Z}\mid\lambda)$, the other major challenges in the case of the HMM (given an observation sequence $\mathbf{Z} = \{Z_1, Z_2, \cdots, Z_T\}$ and a model λ) is computing the state sequence $\mathbf{X} = \{X_1, X_2, \cdots, X_T\}$ that best explains the model. A single best-state sequence is defined as the state sequence X that maximizes $P(\mathbf{X}\mid\mathbf{Z},\lambda)$, which is equivalent to maximizing $P(\mathbf{X}\mid\mathbf{Z},\lambda)$.

The Viterbi algorithm is a dynamic programming-based algorithm used to compute the best-state sequence. Before going into the details of the algorithm, let's define a quantity $\delta_t(i)$ as the best score along a single-state sequence at time t, which accounts for the first t observations and ends in the state s_i. This can be defined as follows:

$$\delta_t(i) = \max_{X_1, X_2, \cdots X_{t-1}} P\left(X_1, X_2, \cdots, X_{t-1}, X_t = s_i, Z_1, \cdots, Z_t \mid \lambda\right)$$

By the induction, we have the following:

$$\delta_{t+1}(i) = \left[\max_i \delta_t(i) a_{ij}\right] \cdot b_{jZ_{t+1}}$$

To actually retrieve a state sequence, we need to keep track of the argument that is maximized for each t and j using the array $\psi_t(j)$. The complete procedure is as follows:

The initialization step:

$$\delta_1(i) = \pi_i \cdot b_{iZ_1} \ \forall i \in [1, N]$$
$$\psi_1(i) = 0$$

The recursion step:

$$\delta_t(j) = \max_{i \in [1,N]} \left[\delta_{t-1}(i) \cdot a_{ij}\right] \cdot b_{jZ_t} \quad 2 \le t \le T$$
$$1 \le j \le N$$

$$\psi_t(j) = \arg\max_{i \in [1,N]} \left[\delta_{t-1}(i) \cdot a_{ij}\right] \quad 2 \le t \le T$$
$$1 \le j \le N$$

The termination step:

$$p^* = \max_{i \in [1,N]} \delta_T(i)$$

$$q_T^* = \arg \max_{i \in [1,N]} \delta_T(i)$$

The state sequence backtracking step:

$$q_t^* = \psi_{t+1}\left(q_{t+1}^*\right), t = T-1, T-2, \cdots, 1$$

This method is similar to what we discussed in the case of forward calculations, except for a few minor changes such as the inclusion of a backtracking step and maximization over previous states instead of summation.

The `hmmlearn` module also facilitates the computation of a state sequence. For example, using the previously defined HMM model with a multinomial emission:

```
# creating a set of random observations
# As the observations can be one of the 5 states that is [0, 4],
# we can create them using np.random.randint
In [24]: random_walk = np.random.randint(low=0, high=5, size=50)

# the array should be in the form of (n_observations, n_features)
# reshaping the array
In [25]: random_walk = random_walk[:, np.newaxis]

# model.decode finds the most likely state sequence corresponding
# to the observation. By default it uses Viterbi algorithm
# it returns 2 parameters, the first one being log probability of
# the maximum likelihood path through the HMM and second being the
# state sequence.
In [26]: logprob, state_sequence = model_multinomial.decode(
                                        random_walk)
```

The next major problem in HMM is to compute the model parameters given the observations. The details of the algorithm are beyond the scope of this book, but we will provide an example of its implementation using `hmmlearn`.

To train an HMM or to compute its model parameters, `hmmlearn` has a fit method in all the HMM classes. The input is a list of the sequence of the observed value. As the **expectation-maximization** (**EM**) algorithm, which is used to compute the model parameters, is a gradient-based optimization method, it will generally get stuck in a local optima. One workaround is to try the fit method with various initializations and select the highest scoring model.

```
In [1]: from __future__ import print_function

In [2]: import datetime
In [3]: import numpy as np
In [4]: import matplotlib.pyplot as plt
In [5]: from matplotlib.finance import quotes_historical_yahoo
In [6]: from matplotlib.dates import YearLocator, MonthLocator,
DateFormatter
In [7]: from hmmlearn.hmm import GaussianHMM


# Downloading the data
In [8]: date1 = datetime.date(1995, 1, 1)  # start date
In [9]: date2 = datetime.date(2012, 1, 6)  # end date


# get quotes from yahoo finance
In [10]: quotes = quotes_historical_yahoo("INTC", date1, date2)


# unpack quotes
In [11]: dates = np.array([q[0] for q in quotes], dtype=int)
In [12]: close_v = np.array([q[2] for q in quotes])
In [13]: volume = np.array([q[5] for q in quotes])[1:]


# take diff of close value
# this makes len(diff) = len(close_t) - 1
# therefore, others quantity also need to be shifted
In [14]: diff = close_v[1:] - close_v[:-1]
In [15]: dates = dates[1:]
In [16]: close_v = close_v[1:]


# pack diff and volume for training
```

```
In [17]: X = np.column_stack([diff, volume])

# Run Gaussian HMM
In [18]: n_components = 5

# make an HMM instance and execute fit
In [19]: model = GaussianHMM(n_components, covariance_type="diag",
                            n_iter=1000)


In [20]: model.fit([X])


# predict the optimal sequence of internal hidden state
In [21]: hidden_states = model.predict(X)


# print trained parameters and plot
In [22]: print("Transition matrix")
In [23]: print(model.transmat_)


In [24]: for i in range(n_components):
             print("%dth hidden state" % i)
             print("mean = ", model.means_[i])
             print("var = ", np.diag(model.covars_[i]))


In [25]: years = YearLocator()    # every year
In [26]: months = MonthLocator()  # every month
In [27]: yearsFmt = DateFormatter('%Y')
In [28]: fig = plt.figure()
In [29]: ax = fig.add_subplot(111)


In [30]: for i in range(n_components):
             # use fancy indexing to plot data in each state
             idx = (hidden_states == i)
             ax.plot_date(dates[idx], close_v[idx], 'o',
                          label="%dth hidden state" % i)
             ax.legend()

# format the ticks
In [31]: ax.xaxis.set_major_locator(years)
In [32]: ax.xaxis.set_major_formatter(yearsFmt)
In [33]: ax.xaxis.set_minor_locator(months)
In [34]: ax.autoscale_view()

# format the coords message box
In [35]: ax.fmt_xdata = DateFormatter('%Y-%m-%d')
```

```
In [36]: ax.fmt_ydata = lambda x: '$%1.2f' % x
In [37]: ax.grid(True)
In [38]: ax.set_xlabel('Year')
In [39]: ax.set_ylabel('Closing Volume')

In [40]: fig.autofmt_xdate()
In [41]: plt.show()
```

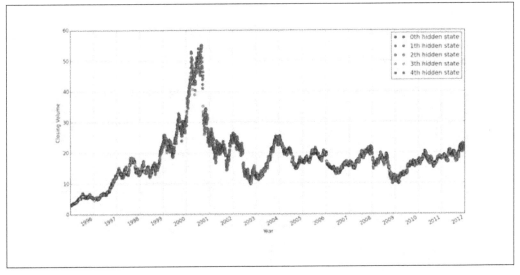

Fig 7.13: Plot showing the closing volume for each of the hidden states across time.
It is the output of the previously stated code.

Applications

One of the major applications of the HMM is in the field of speech recognition. In this section, we will briefly describe the process of speech recognition.

In speech recognition, our job is to compute the most probable word corresponding to a speech signal or acoustic observation. Our aim is to compute the following:

$$\hat{W} = \arg\max_{W \in \mathbf{W}} P\left(W \mid O\right)$$

$$= \arg\max_{W \in \mathbf{W}} \frac{P\left(O \mid W\right) \cdot P\left(W\right)}{P\left(O\right)}$$

$$= \arg\max_{W \in \mathbf{W}} P\left(O \mid W\right) \cdot P\left(W\right)$$

Here, O corresponds to the acoustic observation and W is the set of all possible words. The likelihood $P(O|W)$ is determined by an acoustic model, and the prior $P(W)$ is determined by a language model.

Fig 7.14 shows the architecture of an HMM-based speech recognition system. There are three major components:

- Acoustic model
- Language model
- Pronunciation dictionary

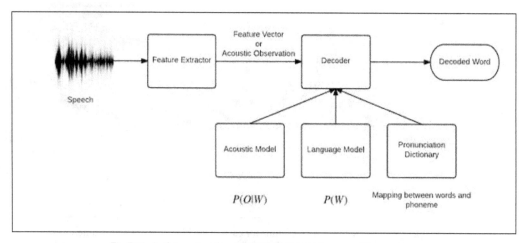

Fig 7.14: Architecture of an HMM-based speech recognition system

The acoustic model

The basic units of sound represented by the acoustic model are the phonetics. For example, the word "bat" is composed of three phonetics, /b/ /ae/ /t/. About 40 such phonetics are required for English. Each spoken letter W can be decomposed into a sequence of K_W base phonetics. This sequence is called its pronunciation. Thus, a word can be represented by an HMM, with hidden state variables being the base phonetics. For example, the HMM for the word *bat* is as follows:

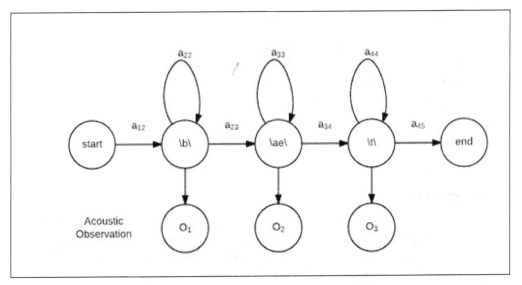

Fig 7.15: An HMM corresponding to the word "bat"

So, with the proper definition of the transition matrix A, the initial state probability distribution π, and the emission probability Θ, we can compute the value of $P(O|W)$ using the forward algorithm, as discussed in the previous sections.

The language model

The language model provides context to distinguish between words and phrases that sound similar. For example, the phrases "recognize speech" and "wreck a nice beach" may be pronounced the same but mean very different things. These ambiguities are easier to resolve when evidence from the language model is incorporated with the pronunciation dictionary and the acoustic model. Further, they also help in faster speech recognition by restricting the search space to the most probable words rather than all possible words. Generally, the N-gram language model is used in most speech recognition applications, where the prior probability of a word sequence $\mathbf{W} = \{W_1, W_2, \cdots, W_K\}$ is computed as follows:

$$P(\mathbf{W}) = \prod_{i=1}^{K} P(W_i \mid W_{i-1}, W_{i-2}, \cdots, W_{i-N+1})$$

Thus, to build speech recognition, we must perform the following steps:

1. For each word υ in the vocabulary, we must build an HMM λ^{υ} by estimating model parameters that optimize the likelihood of the training set acoustic observation for the υth word.

2. Build a language model corresponding to the vocabulary.

3. For each acoustic observation $\mathbf{O} = \{O_1, O_2, \cdots, O_T\}$, we must compute the value of $P(\mathbf{O} \mid \lambda^{\upsilon})$ and select the value of v that maximizes $P(\mathbf{O} \mid \lambda^{\upsilon}) \cdot P(\upsilon)$.

Summary

In this chapter, we discussed special cases in graphical models that are widely used in the real world. We discussed the Naive Bayes model, which is a very simple model but is widely used in text classification and is known to give very good results. Then, we talked about DBNs, which are generally used in cases where we want to model some problem in which the values of the variables change with time. We discussed the Hidden Markov model, which is a very simple case of the DBN and is widely used in the field of speech recognition.

Index

Symbol

0/1 error 162

A

approximate inference
about 207
belief propagation 208, 209
pseudo-moment matching 208, 209
approximate messages
about 117-120
computing 120-122
inference 123
assumptions, dynamic Bayesian
networks (DBNs)
discrete timeline assumption 232
Markov assumption 232, 233

B

Bayesian classifier 218
Bayesian models
about 13, 14
converting, into Markov models 47-49
D-separation 22
factorization, of distribution
over network 16, 17
Markov models, converting into 51, 52
representation 14, 15
Bayesian networks
and Markov network 47
implementing, pgmpy used 17
importance sampling 145, 146
pattern, reasoning 20, 21
representation 18-20
structure learning 183, 184

Bayesian parameter estimation
about 175-177
for Bayesian networks 179-181
local decomposition 183
priors 177, 178
Bayesian score
for Bayesian networks 193, 194
for Markov models 214
belief propagation
about 72, 208, 209
clique tree 72, 73
message passing 76-80
using, for MAP 95, 96
versus variable elimination 100, 101
with approximate messages 117-120
belief update propagation
about 132, 133
MAP inference 133-137
Bethe cluster graph 116

C

causal reasoning 21
chordal graphs 53-55
classification error 162
clique tree
about 72
calibration 80-82
constructing 73-76
defining 73
cluster graph belief propagation 112-114
cluster graphs
Bethe cluster graph 116
constructing 115
constructing, with pairwise Markov
networks 115, 116

collapsed importance sampling 155-157
collapsed particles 138, 154
conditional independence 3, 4
conditional probability
distribution. *See* CPD
constrained satisfaction problem (CSP) 137
constraint-based structure learning
about 184-186
in Markov models 210-212
limitations 212
structure score learning 187
context-specific CPDs
about 28
Rule CPD 30
Tree CPD 28, 29
CPD
about 8, 9, 14, 31, 141
context-specific CPDs 28
deterministic CPDs 26, 27
representations 26
representing, pgmpy used 9, 10

D

decoding 134
deterministic CPDs 26, 27
Directed Acyclic Graph (DAG) 14
directed graphical model 1
discriminative learning
about 165
versus generative training 165
distributions
and graphs, relating 24
graphs, constructing from 46
D-separation
about 22, 45
direct connection 22
indirect connection 22-24
dynamic Bayesian networks (DBNs)
about 231
assumptions 231, 232
model representation 233-235

E

edges 11, 12

energy function
about 106
energy term 107
entropy term 107
exact inference
problem solving 107-110
expectation-maximization (EM)
algorithm 249
expected log-likelihood 161

F

factor 33
factor division
about 83
implementing 84-87
factor graph 42, 43
factor maximization
about 91
example 92
Flat Tyre (F) 28
forward-backward algorithm 243-246
forward sampling 139, 140
full particles 138

G

generative learning
about 165
versus discriminative training 165
Gibbs distribution
and Markov network 38-41
Gibbs sampling
about 148, 149
Markov chain 149-152
gradient ascent 202-206
graphical model
about 1
directed graphical model 1
undirected graphical model 1
graphs
and distributions, relating 24
constructing, from distributions 46
IMAP 24, 25
IMAP, to factorization 25

graph theory
about 11
cycles 13
edges 11, 12
nodes 11, 12
paths 12, 13
trails 13
walk 12, 13

H

Hamming loss 163
Hidden Markov model (HMM)
about 235-238
applications 251
forward-backward algorithm 243-246
HMM-based speech recognition system 252
observation sequence, generating 238-242
probability of observation,
    computing 242, 243
state sequence, computing 247-249
HMM-based speech recognition system
about 252
acoustic model 252, 253
language model 253, 254

I

IMAP
about 24, 25
to factorization 25
importance sampling
about 141-145
in Bayesian networks 145, 146
marginal probabilities, computing 147
normalized likelihood weighting 147
ratio likelihood weighting 147
independence
about 3, 4
representing, pgmpy used 6, 7
**independently and identically distributed
    (IID) 159**
induced graphs
induced width 70
tree width 70
width 70

inference
about 57
belief update propagation 132, 133
complexity 59
example 58, 59
sum-product expectation
    propagation 123-132
with approximate messages 123
IPython
installing 5
URL 5

J

joint probability distribution
about 3
representing, pgmpy used 7, 8
junction tree. *See* **clique tree**

L

Lagrangian multipliers
using 108
Lauritzen-Spiegelhalter algorithm 87
learning task
about 165
data observability 166
density estimation 160-162
empirical risk 164
general ideas 160
goals 160
knowledge discovery 163
model constraints 165
optimization problem 163
overfitting 164
specific probability values,
    predicting 162, 163
Lidstone smoothing 229
likelihood function
about 198, 199
gradient ascent 202-206
log-linear model 200, 201
likelihood score
for Markov models 213
likelihood weighting 141, 142
log-linear model 200, 201

M

MAP
belief propagation, using 95, 96
variable elimination, using 90, 91
MAP inference 89, 90, 133-137
marginal probabilities
computing 147
Markov blanket 45
Markov chain
distributions converge, checking 152
Gibbs sampling 149-152
using 152-154
Markov chain Monte Carlo methods 148
Markovian 232
Markov models
Bayesian models, converting into 47-49
constraint-based structure learning 210
converting, into Bayesian models 51, 52
likelihood score 214
maximum likelihood parameter
estimation 197
score-based structure learning 212, 213
structure learning 210
Markov models, independencies
global independencies 211
local Markov independencies 211
pair-wise independencies 211
Markov network
about 32, 33
and Bayesian networks 47
and Gibbs distribution 38-41
factor operations 35 37
Independencies 44-46
maximum likelihood parameter
estimation 197
parameterizing 33-35
Markov process 235
maximization 91
maximum likelihood parameter estimation
in Markov networks 197
learning, with approximate inference 207
likelihood function 198, 199
score-based structure learning 212, 213
structure learning 210

message passing
about 76-80
implementing, with factor division 83-87
variables from different clusters,
querying 88, 89
with division 82
moral graph 49
moralization, of network 49
most probable assignment
example 96
searching 96
multinomial Naive Bayes model 229
multiple transitioning model 152
multivariate Bernoulli Naive Bayes model
about 224-227
implementation 227
**mutilated network proposal
distribution 145**

N

Naive Bayes model
about 217-219
best model, selecting 231
multinomial Naive Bayes model 229
multivariate Bernoulli Naive
Bayes model 224-227
types 223
usage 220-223
nodes 11, 12
**normalized importance sampling
estimator 145**
normalized likelihood weighting 147

O

optimization problem 104, 105

P

pairwise independency 45
pairwise Markov networks
cluster graphs, constructing 115, 116
parameter learning
about 166
maximum likelihood estimation 166-169

maximum likelihood estimation, for
    Bayesian networks 171-174
maximum likelihood principle 169, 170
particle 138
particle-based methods 138
Perfect Map 25
pgmpy
  installing 5
  URL 6
  used, for implementing Bayesian
      networks 17
  used, for implementing CPD 9, 10
  used, for predicting variable states
      from model 97-100
  used, for representing independence 6, 7
  used, for representing joint probability
      distribution 7, 8
probability theory
  about 2
  conditional independence 3, 4
  independence 3, 4
  random variable 2, 3
propagation-based approximation algorithm
  about 110
  cluster graph belief propagation 112-114
  cluster graphs, constructing 115
  example 111, 112
pseudo max-marginals 134
pseudo-moment matching 208, 209

R

random variable 2, 3
Rao-Blackwellized particles 154
ratio likelihood weighting 147
relative entropy 104
Rule CPD 30

S

sampling-based approximate
    methods 138, 139
score-based structure learning
  about 185
  Bayesian score 214
  in Markov models 212, 213
  likelihood score 213

structure learning
  about 183
  constraint-based structure learning 210-212
  in Bayesian networks 183, 184
  in Markov models 210
  methods 184
structure learning, methods
  Bayesian model averaging 185
  constraint-based structure learning 184
  score-based structure learning 185
structure score learning
  about 187
  Bayesian score 190-193
  likelihood score 187-190
sum-product expectation
    propagation 123, 125, 131, 132

T

target distribution 143
tf-idf 226
tools
  IPython, installing 5
  pgmpy, installing 5
Tree CPD 28, 29
triangulation 53
two-time slice Bayesian network
    (2-TBN) 235

U

undirected graphical model 1
unnormalized importance sampling
    estimator 144

V

variable elimination
  about 60, 62
  analyzing 66-69
  elimination order, searching 69, 70
  example 64, 65
  using, for MAP 90, 91
  versus belief propagation 100, 101
variable elimination order
  cost criteria 71
  searching 69, 70
  searching, chordal graph property used 71

variable elimination order, cost criteria
  min-fill 71
  min-neighbors 71
  min-weight 71
  weighted-min-fill 71
variables connection
  common cause 23
  indirect causal effect 23
  indirect evidential effect 23
vertices 12

W

**weighted importance sampling
      estimator 145**

Thank you for buying
Mastering Probabilistic Graphical Models Using Python

About Packt Publishing

Packt, pronounced 'packed', published its first book, *Mastering phpMyAdmin for Effective MySQL Management*, in April 2004, and subsequently continued to specialize in publishing highly focused books on specific technologies and solutions.

Our books and publications share the experiences of your fellow IT professionals in adapting and customizing today's systems, applications, and frameworks. Our solution-based books give you the knowledge and power to customize the software and technologies you're using to get the job done. Packt books are more specific and less general than the IT books you have seen in the past. Our unique business model allows us to bring you more focused information, giving you more of what you need to know, and less of what you don't.

Packt is a modern yet unique publishing company that focuses on producing quality, cutting-edge books for communities of developers, administrators, and newbies alike. For more information, please visit our website at www.packtpub.com.

About Packt Open Source

In 2010, Packt launched two new brands, Packt Open Source and Packt Enterprise, in order to continue its focus on specialization. This book is part of the Packt Open Source brand, home to books published on software built around open source licenses, and offering information to anybody from advanced developers to budding web designers. The Open Source brand also runs Packt's Open Source Royalty Scheme, by which Packt gives a royalty to each open source project about whose software a book is sold.

Writing for Packt

We welcome all inquiries from people who are interested in authoring. Book proposals should be sent to author@packtpub.com. If your book idea is still at an early stage and you would like to discuss it first before writing a formal book proposal, then please contact us; one of our commissioning editors will get in touch with you.

We're not just looking for published authors; if you have strong technical skills but no writing experience, our experienced editors can help you develop a writing career, or simply get some additional reward for your expertise.

Building Probabilistic Graphical Models with Python

ISBN: 978-1-78328-900-4   Paperback: 172 pages

Solve machine learning problems using probabilistic graphical models implemented in Python with real-world applications

1. Stretch the limits of machine learning by learning how graphical models provide an insight on particular problems, especially in high dimension areas such as image processing and NLP.

2. Solve real-world problems using Python libraries to run inferences using graphical models.

3. A practical, step-by-step guide that introduces readers to representation, inference, and learning using Python libraries best suited to each task.

Expert Python Programming

ISBN: 978-1-84719-494-7   Paperback: 372 pages

Best practices for designing, coding, and distributing your Python software

1. Learn Python development best practices from an expert, with detailed coverage of naming and coding conventions.

2. Apply object-oriented principles, design patterns, and advanced syntax tricks.

3. Manage your code with distributed version control.

Please check **www.PacktPub.com** for information on our titles